Think About This—

* Some chemical substances of known harmfulness, are added to our foods daily.
* Other chemical substances are used with little or no testing of their safety.
* And Federal government agencies seem to be unable to keep track of them all.

The best hope for the consumer is **information**—what these additives are, what they do, what dangers they represent—and, above all, what foods are safest. This is the kind of information Beatrice Trum Hunter has assembled in this book. It will help you select your food wisely.

Other books by Beatrice Trum Hunter

The Natural Foods Cookbook
Consumer Beware!
Gardening Without Poisons
Beatrice Trum Hunter's Favorite Natural Foods
The Natural Foods Primer
Yogurt, Kefir & Other Cultured Milk Products
Fermented Foods and Beverages
Food Additives and Your Health
The Whole-Grain Baking Sampler
The Mirage of Safety
The Great Nutrition Robbery
The Sugar Primer
The Food Connections

BEATRICE TRUM HUNTER'S

Additives Book

Keats Publishing, Inc.　New Canaan, Connecticut

Beatrice Trum Hunter's

FOOD ADDITIVES AND YOUR HEALTH

Portions of this book were first published as articles in
CONSUMER BULLETIN, publication of
Consumers' Research, Inc., Washington, New Jersey 07882

Pivot Original Health Edition published 1972
Revised Edition 1980

Copyright © 1972, 1980 by Beatrice Trum Hunter

All rights reserved

Printed in the United States of America

Library of Congress Catalog Number: 72-83521

PIVOT ORIGINAL HEALTH BOOKS are published by
Keats Publishing, Inc., 36 Grove Street (Box 876),
New Canaan, Connecticut 06840

To F. J. Schlink,
with homage

To the question whether I am a pessimist or an optimist, I answer that my knowledge is pessimistic, but my willing and hoping are optimistic.
—Albert Schweitzer

Contents

Some Facts About Food Additives	1
Types of Intentional Additives and Their Functions	13
A Glossary of Medical Terms	35
Identification of Acronyms	39
Some Additives Commonly Listed on Food Labels	41
What Is to Be Done?	115
Suggested Reading	119
Index	125

Some Facts About Food Additives

"It is a common saying that we are what we eat. If this is true, then Americans are becoming a nation of processed, packaged, and preserved people. Last year, Americans bought more processed than fresh foods for the first time in our history. We spend more than $60 billion for these convenience foods including such items as TV dinners, snack foods of all kinds, and frozen foods.

"With these foods we each consume every year more than four pounds of chemical preservatives, stabilizers, colorings, flavorings and other additives. And the amount of these artificial substances is increasing every year. Their use has doubled in the past 15 years, from 400 million pounds to more than 800 million pounds. Today, more than 3,000 chemicals are deliberately added to our foods.

"These developments raise three basic questions: (1) How much do we know about the hazards to human health from these chemicals? (2) How much assurance of chemical safety should we require? and (3) What must the Federal Government do to assure that the chemicals we absorb are safe?"

These were the opening remarks by Senator Abraham Ribicoff, chairman of the Committee on Government Operations, United States Senate, in hearings in April 1971 on "Chemicals and the Future of Man." The answers to Ribicoff's pointed questions affect the health and welfare of every American in ways not yet fully appreciated by the general public. Chemicals ingested in food and in water, as well as from skin contact, and by inhalation—from the total outer environment—may be causing a new kind of internal pollution to our bodies, and may be seriously damaging to our health. But these chemicals may affect far more than our own health. In addition to exposing our bodies to the threats of chronic toxicity and carcinogenicity, (*see* A Glossary of Medical Terms, pages 35-37) these substances may affect the health and vitality of future generations.

Presently, more than a thousand chemicals added to our foods have never been tested for their potentialities of causing cancer, genetic damage or birth defects. The public wrongly *assumes* that all chemicals contained in foods (and drugs) have been tested for safety and effectiveness. Unfortunately, this is not true. Hundreds of chemicals on a list of substances "Generally Recognized As Safe," known as the GRAS list, and other substances routinely used by the food industry, have never been fully tested for possible harmful effects. If the Federal Food Additive Amendment of 1958 had been vigorously enforced, a majority of the chemical additives on the GRAS list would probably have been excluded.

There has been a loose arrangement between the Food and Drug Administration (FDA) and food processors in the acceptance of food additives for the GRAS list. New additives, introduced after 1958, were to be included on the basis of scientific evaluation of safety. But manufacturers, with FDA's willingness, have reserved for themselves the right to determine whether or not a food additive is safe. Food processors have not needed to file a petition for its use, nor even to inform the FDA of its planned use. Clearly, such a lax procedure could not have resulted if the Federal agency had fulfilled its obligation of vigorously enforcing the law. Because of this lack of law enforcement, it is possible that currently some food additives are being used illegally and without any independent scientific review of their safety. Even when scientific findings have been reported from other quarters regarding the hazards of some food additives on the GRAS list, the FDA has failed to act promptly and decisively. An FDA official even went so far as to state publicly that the agency is not bound to remove a substance from the GRAS list on the basis of initial, but isolated, evidence of potential danger that isn't widely accepted by scientists. "GRAS doesn't have to mean unanimously recognized as safe," he added. Such an interpretation fails to offer consumer protection by the agency charged with this responsibility.

This looseness of official policy regarding the GRAS list was described by Senator George McGovern as "the never-never land of nonregulation." Once food additives are on the GRAS list, they automatically enjoy certain privileges. They are exempt from the legal definition of a food additive, and are not subject to regulations concerning food additives as long as they remain on the GRAS list. In order for a substance to be removed, the FDA must demonstrate the harmfulness of the additive. Clearly, the wiser policy would be to pretest every food additive before it is approved for use, and to establish a "proved as safe" list. Such a procedure would result in a far smaller list, and the consuming public would be more adequately protected.

The federal regulatory wheels grind slowly. In 1973, the FDA began to implement a 1969 presidential mandate to review the entire GRAS list. The review, finally launched, came under critical attack. The total study had been estimated to cost taxpayers some $20 million. By the end of the first year, with the project hardly begun, the cost of the review had already exceeded the projected sum. Close identification of the National Academy of Sciences/National Research Council's Food Protection

Committee with industrial interests made this group singularly inappropriate as a major source of "independent advice" for the review. The task was transferred to the Federation of American Societies for Experimental Biology (FASEB), a group that regarded the review as "a new and unusual endeavor."

The review has consisted largely of sifting through old, inadequate data, much of it performed by industry, or its privately contracted laboratories. The review has relied on relatively few new studies conducted with more sophisticated testing methods or techniques.

Meanwhile, as more safety questions about food additives gained prominence in the news—including saccharin, nitrite, acrylonitrile, vinyl chloride, and others—the FDA took a broader view. In 1977, the agency announced that it would conduct a periodic review of *all* food additives, not only those on the GRAS list, to assure their safety by modern standards. Under this proposal, all substances added to foods, including preservatives, colors, flavors, and components from packaging that may migrate into food and interact with it, would all undergo regular scientific reevaluation. Another important aspect to be included is an industry-wide survey to learn how much of each additive an average individual may be consuming.

"How much do we know about the hazards to human health from these chemicals?" asked Senator Ribicoff. Analyzing a chemical food additive is not merely a problem of exploring its possible acute harmfulness. Yet past testing techniques were developed largely for the study of direct toxic effects, and the "safety" of some additives currently in use is based on what we presently understand to be insufficient criteria. Scientists have become aware of the need to study the untoward results, those which are slight, unnoticed, delayed and indirect. These are the subtle effects on the human system at the basic cellular level, resulting from hundreds or even thousands of substances biologically foreign to the body being consumed daily in common foodstuffs, over many years, or even during an entire lifetime. The problem is complicated by the need to examine chemical food additives in relation to other chemical and physical environmental factors, such as heavy metals, pesticide residue, tobacco, drugs, alcohol and radiation.

Some food additives produce chemical changes in the food itself by altering its biological structure. Others may combine with constituents naturally present in the food to form new toxic compounds. Others may produce derangements in the human system and be so insidious that their presence does not become apparent until long after the original exposure. Because of the time lag, they may not even be suspected as the instigators of harm.

Until now, most of the proof for safety of food additives has been established through animal tests. Even with carefully devised tests, it is difficult to extrapolate results in terms of human beings. Animals do not necessarily react to substances in the same manner as people. For examples: Beta-naphthylamine, one of the most active human cancer-inciting chemicals, failed repeatedly in tests to induce tumors in animals. Many years passed before a species was found whose response to this substance was similar to man's. Or, consider meclizine, a drug used for pregnancy sickness. It is teratogenic in rats, but apparently not in the few women who have taken it. However, with thalidomide, women are sixty times more sensitive than mice, one hundred times more sensitive than rats, two hundred times more sensitive than dogs, and seven hundred times more sensitive than hamsters.

The Joint Food and Agricultural Organization of the World Health Organization (FAO/WHO) Expert Committee on Food Additives has pointed out the unreliability of animal tests. It has cautioned that absolute proof of nontoxicity *cannot* be established through animal tests because of the special conditions of human life and individual idiosyncrasies. Experimental animals are tested in a thoroughly controlled environment, impossible with human beings. In addition, the old, the ill, the allergic, or the very young human may be far more sensitive to substances than the well-fed rat that has a diet balanced in all known nutrients, and is isolated from other chemicals in the environment. Pharmacologically, the fetus or newborn child is an entirely different organism from the adult human. But animal testing is not geared to meet these special requirements.

Like animal testing, laboratory analyses of chemical food additives are inadequate measures of safety. Dr. Arnold J. Lehman, former director of the Pharmacological and Toxicological Divisions of the FDA, admitted a present lack of precise knowledge regarding the ultimate fate of chemicals in the body. For a long time, he reported, toxicologists thought that the body was able to call upon some special mechanism to detoxify poisonous substances absorbed in the body. Experts had also believed that the toxic substances were broken down into less toxic compounds. More recent findings demonstrate the error of both beliefs. Special mechanisms do *not* exist. Toxic substances are subject to the metabolic processes that function normally in the body. Absorbed products are not necessarily less harmful than the original materials. In some instances, the toxicity of the breakdown product has been found to be greater than that of the original. This point was well illustrated with cyclohexylamine and dicyclohexylamine, two breakdown products from cyclamate.

Toxicologic studies of chemical food additives are complicated further by a mechanism known as synergism. The toxicity of one chemical can be increased considerably by an interaction with another one that may be present. During a given meal, the consumer may eat a number of chemical additives. It is impossible to know the potential harm in the infinite variety of chance combinations that may occur. For this reason alone, it would be prudent to keep the list of officially approved additives small; it is impossible to test for all the innumerable synergistic effects.

A few synergistic effects have already been found, not only among food additives but among pesticides, drugs, alcohol and other substances. An important example is the consumption of alcohol before exposure to carbon tetrachloride fumes. Another is the sometimes fatal combination of alcohol and barbiturates. Radiation experts say that exposure to radiation, combined with cigarette smoking increases the potential damage tenfold. Data suggests that asbestos workers who smoke cigarettes are ninety times more likely to die of lung cancer than persons who neither work with asbestos nor smoke.

Chemicals, including food additives, that can trigger cancer are of two kinds. One kind, the carcinogens, are direct cancer-inciters. The second kind, known as weak carcinogens or cocarcinogens, is normally not cancer-inciting. But indirectly through synergism, or in combination with other substances, weak carcinogens or cocarcinogens *can* become active carcinogens. Some chemical food additives, such as Food, Drug and Cosmetic (FD&C) Citrus Red No. 2, used to color the skins of citrus fruit, are weak carcinogens.

Toxicity tests do not necessarily show cancer-inciting properties of a chemical. As a rule, the smallest effective carcinogenic dose is lower than the toxicity threshold. Frequently, it is even within a nontoxic level of the smallest acute or chronic toxic dose of a chemical. In addition, there is no direct relationship between the relative toxic potential of a chemical and its carcinogenic property. Benzene, for instance, is highly toxic to the blood-forming tissue, but it has only a relatively mild cancer-inciting effect. On the other hand, beta-naphthylamine, which has a very low degree of toxicity for man, is one of the most potent of human carcinogens. In fact, there appears to be an antagonism between toxic and carcinogenic properties of chemicals. Highly toxic chemicals are apt to kill cells outright rather than merely injure them or modify them biologically. Because of this immediate destruction, the cells do not survive long enough to become abnormal and cancerous. These facts are important, because many tests for chemical food additives, especially in the past, were for toxicity, not for carcinogenicity (nor teratogenicity, nor mutagenicity).

"How much assurance of chemical safety should we require?" asked Senator Ribicoff. Far more than we are presently demanding. Animal tests are extraordinarily insensitive. According to Dr. Samuel S. Epstein, then Chief of the Laboratories of Carcinogenesis and Toxicology, Applied Microbiology and Histology, at the Children's Cancer Research Foundation in Boston, "Let us say we introduce into the environment a food additive which produces cancer or birth defects in one in every 10,000 men or women. Let's also assume that the sensitivity of rats to this carcinogen in particular is the same as that of man. Therefore, you would need test groups of at least 10,000 rats or 10,000 mice to get one cancer or one birth defect. In practice, we test say 50 rats or 50 mice per dose level per chemical. Under these situations, if we test chemicals at levels of human exposure our possibility of picking up a weak carcinogen or teratogenic effect is very low."

In order to detect possible low incidences of carcinogenic properties, "mega-mouse" experiments have been suggested, using as many as one hundred thousand mice per experiment. A single mega-mouse experiment is estimated to cost about $15 million, and would still leave most problems unsolved.

Instead, Dr. Epstein and others suggest the wisdom of testing at higher dose levels, up to maximally tolerated dose levels. Under these circumstances, one can perhaps reduce the gross insensitivity of animal tests. However, it is still difficult to extrapolate. In some instances man is more sensitive, and in some instances less. Comprehensive safety testing demands the availability of detailed chemical and comparative metabolic data on the food additive under scrutiny. The requirements for such data may make it necessary to conduct carefully controlled metabolic studies in human volunteers, as was done with cyclamates, after safety has been established in comprehensive animal tests. The food additive must be given in tests that will show the acute, subacute and chronic toxicities, and that will trace the metabolic fate of the substance.

Can a safe level be established for a food additive that has been shown to be cancer-inciting? This question continues to raise a storm of controversy. It began in 1958, when the Delaney Clause was written into the Federal Food, Drug and Cosmetic Act Amendment. The Delaney Clause provided that "no additive shall be deemed safe if it is found to induce cancer when ingested by man or animal, or if it is found, after tests which are appropriate for the evaluation of the safety of food additives, to induce cancer in man or animal." Over strong industry protests, the Delaney Clause was retained in 1958, and also in the Color Additive Act in 1960. The clause made it mandatory for the FDA to withdraw from the market aminotriazole-tainted cranberries, stilbestrol-contaminated chickens, and more recently, cyclamates, for all these substances showed cancer-inciting properties in test animals. Because of the great emotional impact of the subject of cancer, realistic industry spokesmen recognized the obstacles for outright appeal of this clause. Instead, efforts have been made to circumvent it. Attempts have been made repeatedly to establish so-called safe levels for carcinogenic substances, including food additives.

Dr. Epstein, as well as many distinguished cancerologists, state emphatically that it *is not* possible to predict safe levels of carcinogens based on an arbitrary fraction of the lowest effective animal dose in a particular experimental situation. They also state that testing at high dosages does *not* produce false positive carcinogenic results. There is no basis whatsoever for an erroneous and often repeated statement that all chemicals are carcinogenic or mutagenic at high doses. It is simply not true.

As yet, there is no mandatory requirement for mutagenic testing of food additives. Such testing is at the discretion of petition reviewers. In recent years, three different expert committees—the FDA Protocol Committee, the Advisory Panel on Mutagenicity to the Mrak Committee, and a National Institute of Environmental Health Sciences task force—have unanimously recommended that mutagenic testing be made mandatory for food additives (and pesticides). Teratogenic effects should also be examined.

"What must the Federal Government do to assure that the chemicals we absorb are safe?" asked Senator Ribicoff. Far more than is presently being done. Both the U.S. Department of Agriculture (USDA) and the FDA have shown carelessness, negligence and indifference to the health and welfare of the consuming public. Both agencies have been slow, dilatory and confused in enforcement of their own policies and regulations.

In addition to an adequate testing program for food additives, and a withdrawal of the harmful ones, there is critical need for public access to scientific data filed with the FDA and the USDA. All formal discussions between agencies, industry and expert governmental and non-governmental committees on all issues relating to human safety and environmental quality belong in the public domain. They should be a matter of open record. Such data should be at all times, and immediately, available to the scientific community as well as the other concerned groups. Consumers and occupational and environmental groups should be adequately represented at the earliest stages of any formal discussions, and should participate in policy decisions regarding proposals for the introduction of technological innovations such as new food additives and other chemicals into the environment. Public access to such data would not only serve consumers, but, in the long range, serve industry as well. Such a policy would give industry far clearer guidelines for testing than are presently being offered. It could prevent the development and marketing of inadequately tested substances, such as nitrilotriacetic acid (NTA) detergents and cyclamates, and in turn prevent the chaos and financial losses that result when these substances are banned.

Types of Intentional Additives and Their Functions

Acidifiers: See Acidulants.

Acidifying agents: See Acidulants.

Acidulants: Additives that provide, or change the degree of acidity in processed foods are called acidulants. The major role of acidulants is to provide an appealing tartness in such foods and beverages as candies, jams, gelatin desserts, sherbets, carbonated soft drinks, instant soft-drink mixes and fruit juices. Acidulants also act as flavoring agents by intensifying certain tastes, blending unrelated taste characteristics, and masking undesirable aftertastes. They act as buffers, by controlling the pH (acid-alkalinity value) of food during different stages of processing, and as a finished product. They act as preservatives, by preventing the growth of microorganisms and the germination of spores that spoil food and thus cause food poisoning and disease. They act as synergists to antioxidants, by preventing food from turning brown or becoming rancid. They act as viscosity modifiers for food products such as cheese spreads and mixtures used to make hard candy. They act as curing agents in combination with other curing components, to increase color, flavor and preservative action. They act on leavening agents in baked goods to release the gas that causes the dough to rise. Acidulants are also called acidifiers and acidifying agents. See also Buffers and Neutralizers.

Adjuvants, flavoring: Substances that facilitate or modify the action of the principal flavoring ingredient or ingredients are known as adjuvants. See also Flavorings, food.

Aerating agents: Combustion-product gases are used with foamed or sprayed food products in pressurized cans. Aerating agents are also called whipping agents. See also Propellants.

Alkalies: Additives that provide or change the degree of alkalinity in processed foods are called alkalies. Alkalies control the pH of foods, which in turn improves the flavor and increases the stability during storage. Alkalies are used to neutralize sour cream when it is used to make butter; to control the pH of canned peas, tomato soup and tomato paste; to reduce the excessive acidity in wine; and to denude tripe. Since alkalies can modify starch, they are sometimes used as glazing agents for pretzels. Alkalies are used also in processing chocolate; in processing olives before they are canned; in extracting color from annatto seeds; and in solutions to peel tubers such as potatoes and fruits such as tomatoes. Alkalies are used in confections, cookies and crackers, baking powder, creamed cottage cheese, ice cream, ice cream syrups, ices, sherbets, prepared pancake mixes, biscuit and muffin mixes, self-rising flours, bleached flour, and in the bleaching of milk for certain cheeses. Alkalies are also called pH adjusting agents. See also Buffers and Neutralizers.

Antibiotics: The preservative action of antibiotics in food is antimicrobial. Antibiotics delay bacterial spoilage of perishable foods during storage, transportation and marketing. They prevent microbial buildups during preparation for other processings such as freezing. Antibiotics are also used as disease-preventives in meat-producing animals, poultry and fish. This use is under current review because of the potential hazards of these drugs. The FDA proposed in 1966 that the use of antibiotics to preserve chilled, uncooked poultry and seafood be prohibited, partly on the grounds that the antibiotics are only "a substitute for good manufacturing practices."

Antibrowning agents: When fruits and vegetables are peeled, cut or sliced, the oxygen present in the air acts upon enzymes present in the food, and the produce discolors. Antibrowning agents inhibit this discoloration.

Anticaking agents: A number of compounds are added to crystalline foods and food powders, such as table salt, baking powder, vanilla powder, garlic salt, malted milk powder and nondairy coffee creamers, to prevent them from absorbing moisture and caking (*see* pages 97-99).

Antifirming agents: See Antistaling agents.

Antifoaming agents: See Defoaming agents.

Antimold agents: Compounds are used to inhibit the molding of many foods, including bread and other baked products (*see* pages 52-53), cheeses, chocolate syrups, jellies, cakes and dried fruits. These compounds are also called mold inhibitors, mold retarders or antimycotic agents. See also Preservatives.

Antimycotic agents: See Antimold agents and Antirope agents.

Antioxidants: The oxidative deterioration of the flavor and odor of fats and fatty constituents of foods is retarded by the use of antioxidants. These compounds also minimize the oxidative destruction of vitamins and essential fatty acids. By delaying rancidity, antioxidants retard or prevent the spoilage of many foods. Although natural antioxidants are found in tocopherols in vegetable oils, most of the antioxidants used in foods are synthetic. Some are petroleum products. At first, synthetic antioxidants were used with fat, but later they were incorporated so extensively into other items that currently they are found in a wide array of factory-processed foods and food packaging materials. They are used in animal fats such as lard, beef tallow, bacon, chicken fat, butter, cream, shortenings and grease; fried foods such as potato chips, doughnuts, and processed meats and fish; baked goods such as crackers, pastries, cakes, candies and cookies; vegetable fats such as shortening, margarine, peanut butter, and salad oils and dressings; ground grain meals and grain germs; a miscellany of other food items, including nutmeats, raisins, milk, candied fruit, whipped topping mixes, imitation fruit drinks, breakfast foods, yeast, extracts, essential oils, spices and pet food; food packaging paper and containers, such as cartons for milk; containers for cottage cheese, ice cream, potato chips, cereals, cookies and pastries; wrappers for breads, butter and cheeses; rubber gaskets that seal food jars; and household wax paper, They are also found in some beverages, chewing gum, cosmetics, drugs and animal feed.

Synthetic antioxidants are of dubious safety. Most of them are known to be toxic to some degree, while some are suspected as possible cancer-inciters (*see* pages 43-44 and 44-46). Adverse findings have been reported, both with laboratory animals and in clinical cases of human sensitivity.

Antioxidants are also known as oxygen interceptors and freshness preservers. See also Preservatives.

Antirope agents: Rope bacteria can form a slimy condition in bread dough. To prevent this, bakers formerly added vinegar. Presently, antirope agents are used. Antirope agents are also called rope inhibitors, or antimycotic agents.

Antistaling agents: The staling of breads, cakes, doughnuts and other baked goods is inhibited by antistaling agents. These additives also help to produce desired volume and texture in baked goods. Antistaling agents are also known as antifirming agents, crumb-firming agents and, incorrectly, as bread emulsifiers. See also Dough conditioners.

Antisticking agents: See Release agents.

Artificial colors: See Colors, food.

Artificial sweeteners: See Nonnutritive sweeteners.

Base components, chewing gum: At last count, twenty-five synthetic and eighteen natural substances were officially permitted as base components of chewing gum. Among the synthetic masticatory substances are petroleum derivatives. Products of petroleum "cracking" are strongly suspected as cancer-inciters.

Binders: Starches and starch-containing products are used in processed meat products such as scrapple and sausages to maintain body and keep the meat, fat and juices together during processing and storing. The amount of starch varies widely, depending on the type of product and official regulations on the local and national levels. Binders are also used in extruded pet food and snack foods. Common binders include cereal flours, cracker and bread crumbs, dry milk powder, casein, potato flour and soy flour.

Bleaching agents: Bleaches are used by several food industries, but they are principally important for flour milling, bread baking and cheese making.

Freshly-milled wheat flour has a yellowish color from small quantities of natural pigments in the grain. When flour is stored for several months it gradually becomes whiter due to oxidation. In 1915 it was found that the bleaching process could be hastened by treating flours with chemicals in the form of gas. These bleaching agents change the yellow pigments to white, and some, in addition, modify the gluten characteristics so as to improve the baking results.

For about twenty-five years, one flour bleach, nitrogen trichloride (Agene), was used. In 1946, Sir Edward Mellanby, a British physician, found that dogs fed large quantities of flour treated with this compound, contracted running fits or canine hysteria. The treated flour also produced convulsions in cats, rabbits, and mink. The toxic compound was later identified as methionine sulfoxamine, which is formed by the action of nitrogen trichloride on methionine present in the flour. In 1949, the FDA prohibited nitrogen trichloride as a flour bleach. The incident is an illustration of how toxic substances may be produced when food additives combine with constituents already present in the food.

Bleaching agents are used by the dairy industry in the production of certain cheeses. Blue and Gorgonzola are relatively white when they are made traditionally from goat's or sheep's milk. When they are made from cow's milk their color is more yellow. Acceptance on the market depends on their being bleached. Bleaching agents are also used to neutralize color that may be present naturally in fats and oils. See also Bread improvers, Dough conditioners and Maturing agents.

Bodying agents: Special thickeners are used with gelatin and edible gums to achieve certain textures. Also called Thickeners.

Bread emulsifiers: See Antistaling agents.

Bread improvers: In addition to bleaches and maturing agents used by flour millers, numerous types of chemical additives are used by bakers. These include yeast foods, dough conditioners, leavening agents, anticaking agents, hardening agents, clarifiers, chill-proofing agents, propellants for products packed under pressure in dispensing cans, drying agents, antifoaming agents and antimold agents. See also Bleaching agents, Dough conditioners and Maturing agents.

Buffering agents: See Buffers.

Buffers: The acidity or alkalinity of foods and beverages is controlled by neutralizing them within limits of good manufacturing practice. The sodium salts of acids are used to control the degree of acidity in soft drinks. The concentrations of acids and buffers used are essentially the same as the level at which these substances occur naturally in fruits. Buffers are used in many processed foods, including corn and table syrups; breakfast cereals; confections; beverages; ice cream; ices; baked goods; jellies and jams; leavening agents; bakery products; prepared pancake, muffin, cake and biscuit mixes; self-rising and phosphated flours; canned potatoes, green and red sweet peppers and tomatoes; frozen dairy products; chocolate products; evaporated milk; dessert mixes; whipped toppings; enriched farina and macaroni; pumping pickle for cured hams; cured and chopped hams; fresh beef blood; cheeses; and frozen fruit drinks. Buffers are used also to retain the carbonation in beverages.

Acids are harmful to the teeth, for they erode the enamel. Buffers can also interfere with normal digestion.

Buffers are also called buffering agents. See also Acidulants and Alkalies.

Carriers: Substances which serve as diluents for others are called carriers. Some carriers are used with bleaching agents of flour.

Certified color additives: See Colors, food.

Certified food colors: See Colors, food.

Chelating agents: See Sequestrants.

Chelators: See Sequestrants.

Chewing gum base components: See Base components, chewing gum.

Clarifiers: See Clarifying agents.

Clarifying agents: Substances are used to filter liquids such as wine, beer and vinegar. Clarifying agents are also called clarifiers and fining agents.

Clouding agents: High-density oils are blended with low-density essential oils to make them easier to emulsify.

Coal tar dyes: See Colors, food.

Coating agents: Compounds are officially sanctioned for use to coat the skins of certain fruits and vegetables to minimize bruises in transport, to prevent drying out or molding. Many of these compounds are toxic.

Starches and modified starches are used to coat many processed foods, including nuts, chocolate confections, hard-sugar confections and chewing gum. French fried potatoes may first be dipped in a starch solution and then fried. The starch coats the surface, reduces oil penetration and yields a crisp product.

Colors, food: Until the middle of the nineteenth century all colors added to foods were of natural origin. In 1856 the first synthetic coal tar dyes were made. They now predominate. More than ninety percent of all food colors presently used are synthetic. Some natural food colors continue to be used because no approved synthetic color has been found as a substitute. Or, because some food companies have found that their customers have an overwhelming preference for food colors of natural origin.

Synthetic food colors are preferred by the food industry because such colors generally have greater coloring power than natural food colors, are more uniform and stable, and are usually cheaper than natural food colors. Food colors are added to thousands of foods, including soft drinks, fruit juices and fruit drinks, cordials, ice cream, ices, sherbet, gelatin desserts and puddings, maraschino cherries, meat casings, prepared mixes, cheeses, butter, margarine, breakfast cereals, candies, jellies and pet food. Without the addition of coloring matter, these products could not be manufactured in the finished form that consumers have been conditioned to accept.

When certain foods are processed, color changes take place. Food colors "correct" these changes so that the finished food products are consistent and have the appearance that is expected.

The history of food color safety is *not* reassuring to consumers (*see* pages 58-69). A number of colors officially sanctioned for use, after adequate testing, have been found to be hazardous and have been banned. Among the nine United States certified food colors currently in use, some have demonstrated adverse effects on laboratory animals, and are under current review. Allergists have reported patients sensitive to some of them. Food colors are also called food dyes, certified food colors, certified color additives, FD&C colors, and FD&C lakes.

Crisping agents: See Firming agents.

Crystalline inhibitors: Substances are added to cooking and salad oils to keep crystals from forming and to keep the oil clear in appearance. Such crystals are harmless.

Crystallization modifiers: Soft sugar-based confections stay soft because the sugar is only partially crystallized. Crystallization modifiers are added to sugar-based confections to extend the shelf life of the candy by permitting it to retain its texture and desirable chewing properties.

Crumb-firming agents: See Antistaling agents.

Curing agents: Substances are used to hasten and stabilize color fixing in cured meats, to modify their flavors, to preserve them and to aid in retaining their juices when they are cooked. See also Fixatives.

Defoaming agents: A number of compounds reduce the foam in cooking, fermenting, bottle-filling and in other processings with fruit juice, wine, beer and other liquids. Some of the officially sanctioned defoaming agents are toxic in large quantities. Defoaming agents are also called antifoaming agents. See Surfactants.

Dough conditioners: Compounds are used by bakers to help produce a dough that is drier, more extensible and easier to machine. Dough conditioners permit the production of a more uniform bread despite variations in flour characteristics and in the amount of mixing. In addition, they give bread some of the properties of antistaling agents: a more uniform grain, a more palatable crumb and a longer shelf life. See also Antistaling agents, Bleaching agents, Bread improvers and Maturing agents.

Dusting agents: See Release agents.

Dyes, food: See Colors, food.

Emulsifiers: Substances that permit the dispersion of tiny particles or globules of one liquid in another liquid are known as emulsifiers. Natural emulsifiers are present in some foods, notably the lecithin in eggs and soybeans. Formerly, the food industry relied on the natural emulsifiers. The lecithin in egg yolk, for example, was the original emulsifier in mayonnaise and still is the major one in this food product.

Synthetic emulsifiers were first introduced for food use in 1928. They now account for most of the food emulsifiers used by processors, for a variety of purposes. These compounds combine substances such as oil and vinegar in salad dressings, and permit the ingredients to stay combined long after mixing ends. They may be used as substitutes for more expensive ingredients such as eggs, milk or butter. When they are added to shortening, for example, cakes can be baked with a higher ratio of sugar to flour, at a lower cost, and without a loss of volume. They are used to improve the texture of cake mixes, and to give ice cream a smoother, more uniform texture, a texture that has been traditionally achieved with cream and milk. They improve bread and roll characteristics such as volume, uniformity and fineness of grain. Batters and dough handle more readily when emulsifiers are added to baking formulas. Emulsifiers extend the shelf life and give homogeneity to some candies and confections. Chocolate, for example, tends to "bloom" or change its surface color if it is exposed to temperature changes because the cocoa butter separates from the chocolate. An emulsifier helps to prevent this condition by keeping the fat in a more stable emulsion within the chocolate. Emulsifiers are used in nondairy creamers (coffee whiteners) to speed the dispersal of the product in coffee. Emulsifiers are used in pressure-dispensed dairy products. They also disperse flavor oils in sherbet, dill oil in pickles and flavorings in soft drinks.

The safety of some of these synthetic emulsifiers is dubious (see pages 53-58 and 100-102). Emulsifiers are also called emulsifying agents, surface active agents, surfactants, thickeners, viscosity agents and wetting agents. See also Stabilizers.

Emulsifying agents: See Emulsifiers.

Enzymes: Enzymes are used as food additives for a variety of purposes. Some tenderize meat tissues. Others are used with baked goods to accelerate the hydrolysis of starch and glucogen. Others are used for curd production in cheese making and to eliminate residues of peroxide used as a bleach.

Extenders: Substances are added to processed foods, especially as diluents or modifiers. Extenders are usually less expensive than the main ingredients. Frequently they consist of soybean, starch, glucose, breading or milk casein. See also Fillers.

Fillers: Substances are added to processed foods, especially to increase the bulk or weight of products, or to dilute expensive materials. See also Extenders.

Fining agents: See Clarifying agents.

Firming agents: To keep the texture of processed fruits and vegetables from becoming soft, compounds are added to canned potatoes, tomatoes, lima beans, peppers and apple slices. Firming agents are also used with maraschino cherries and pickles, to keep them crisp. Firming agents also aid the coagulation of certain cheeses. Firming agents are also called crisping agents.

Fixatives: Color fixatives are used to keep meat and processed meat red in appearance (*see* pages 102-106). Fixatives are also added to flavorings in order to prevent the more volatile ingredients from evaporating.

Flavor enhancers: In the low concentrations used, flavor enhancers usually add no flavor of their own to foods, but heighten or modify the flavor of specific food ingredients already present. Certain flavor enhancers create a sense of viscosity in liquid foods greater than is present in reality. They give soups more body, or what the trade terms "greater mouthfulness" or "mouth feel." Some flavor enhancers accentuate meatiness and are used in meat products, canned vegetables, sauces, gravies, bouillons and soup bases. In the United States, these additives have been especially promoted to the trade as a means of reducing the amount of natural beef extract required in foods, or of replacing it entirely. Beef extract, a byproduct in the processing of corned beef, is becoming scarce and is considerably more expensive than the flavor enhancers that replace it.

Flavor enhancers are used with fruit flavors, to intensify the flavors of foods high in carbohydrates. These foods include soft drinks, fruit drinks, jams and gelatin desserts. Certain flavor enhancers commonly intensify strawberry and raspberry flavors. They are also used with pineapple, orange, black cherry and other flavors.

Some flavor enhancers have synergistic properties. When two or more of these compounds are used together, the flavor potentiation increases enormously. Some compounds are so powerful that they can intensify the flavor of foods when the enhancers are used in amounts of parts per billion or less. Flavor enhancers are also called flavor intensifiers, flavor-modifying compounds and flavor potentiators. One flavor enhancer is under current review (*see* pages 81-91).

Flavor enzymes: Many fresh foods contain volatile flavor compounds. During processing, such as blanching, dehydration or freeze drying, all or part of these volatile compounds are lost and the flavor declines. By treating the flavorless processed food with appropriate enzymes, the flavor precursors may be converted to flavor compounds, thus improving the flavor of the food. This does not necessarily restore nutritional losses from the processing. The principle of enzymatic flavor restoration can be applied to most processed foods.

Flavor intensifiers: See Flavor enhancers.

Flavor modifiers: See Flavor enhancers.

Flavor potentiators: See Flavor enhancers.

Flavorings, artificial: It is possible to have a flavor that contains all natural ingredients, but this flavor is considered artificial if it has no counterpart in nature. See also Flavorings, food.

Flavorings, food: Flavorings are added to foods either to produce or to modify flavor. They can give food the flavor of the flavoring, supplement or modify its own flavor, or mask the original flavor of the material. Flavorings are commonly added to foods that lose their original flavorings as a result of processing. They are used in processed meats, baked goods, ice cream, ices, candy, fruit-flavored toppings, fruit-flavored gelatins, puddings, chewing gum, fruit syrups, beverages, condiments, pickles, cottage cheese, shortening and liquor.

Flavoring substances and adjuvants form the largest single category of food additives. More than a thousand different flavoring substances are used. With the enormous increase in the sale of "convenience foods" the total amount of flavorings and flavor enhancers has risen 92 percent in a decade.

Most food flavorings presently used are synthetic. The reasons are numerous and varied. The cost of producing many natural flavorings has reached the point where they are prohibitive. They are not uniform in composition, nor are they stable. They have a fluctuating availability, and they are in short supply. Industry estimates that the world's total supply of natural vanilla flavoring actually would be inadequate to flavor just the amount of vanilla ice cream eaten in the United States—without counting all of the frozen desserts, baked goods, candies and beverages also flavored with vanilla. Similarly, it has been claimed that if strawberry-flavored foods were produced only from actual strawberries grown in the country, the entire annual supply would be used in a few months by a city the size of Syracuse, New York.

The history of the safety of food flavorings is no more reassuring to consumers than that of food color safety, whether the flavorings are from natural or synthetic sources (*see* pages 69-73). See also Adjuvants, flavorings; Flavor enhancers; Flavor enzymes; Flavorings, artificial; Flavorings, imitation; and Oleoresins.

Flavorings, imitation: Imitation flavorings contain all, or some portion of, nonnatural materials. For example, unless an orange flavoring is made entirely from orange, it is considered imitation See also Flavorings, food.

Flavorings, natural: See Flavorings, food.

Flavorings, synthetic: See Flavorings, food.

FD&C colors: See Colors, food.

FD&C lakes: Water-soluble FD&C colors, prepared by extending the aluminum or calcium salt of the color on a substratum of alumina, have been permitted for use in coloring nonaqueous food products since 1959. Each lake is considered to be a straight color and is listed under the name of the color from which it is formed.

Foaming agents: See Aerating agents.

Food colors: See Colors, food.

Food dyes: See Colors, food.

Freshness preservers: See Antioxidants.

Fumigants: Toxic gases are used to control pests in soil or on crops after they are harvested, in granaries, in mills, in warehouses, in ships and freight cars. These fumigants are used on dried grains, beans, peas and lentils; dried fruit; and nuts. They are used in extracting oils from nuts and seeds. Fumigants are toxic materials.

Fungicides: Undesirable fungus spores, which form on or in certain foods, are destroyed by fungicides.

Glazing agents: Substances are used in order to form a hard surface especially to coat candy. See also Coating agents.

Pan glazes are used by confectioners and bakers to prevent candies and baked goods from sticking to the pans. See also Release agents.

Humectants: Certain foods are unacceptable unless they retain enough moisture to keep soft and pliable. Humectants help retain moisture and prevent drying and undesirable brittleness in shredded coconut, marshmallows, jelly-like candies, pastilles, chocolates and ice cream. Humectants are also called hydroscopic agents, moisture-retaining agents and water-retaining agents.

Hydroscopic agents: See Humectants.

Leavening agents: Substances used to make foods light in texture are called leavening agents. Although air and steam sometimes act as leaveners, by far the most important ones are yeast, baking powder and baking soda. Before the chemical reaction of leavening was understood, yeast was used almost exclusively. Today's chemically-released carbon dioxide leaveners are tailored to meet specific needs. In commercial bakeries, where large batters are held for varying lengths of time, or even refrigerated overnight, slow-acting baking powders are used. In the field of preleavened mixes such as cakes, pancakes, muffins, biscuits and self-rising flour, special combinations are provided to produce the desired results. In such mixes, the stability of the leavening is vital. For this purpose, special prepared acid salts are used.

For sour milk biscuits, phosphated flour is used. Such flour contains acid salts to neutralize the excess alkalinity that results when the milk lacks enough acidity to react with all the sodium bicarbonate used. Leavening agents are also called leaveners. See also Dough conditioners.

Lubricating agents: See Release agents.

Maturing agents: When flour is stored and allowed to age for several months it "matures," which makes it more satisfactory for baking. For centuries this natural aging was the only way millers and bakers could produce the desirable quality for baked goods. A few years after bleaches were introduced into flour (1915), it was discovered that some of the bleaches, as well as other chemicals, could speed up the aging or "maturing" process in flour. The maturing agents modify the gluten characteristics in the flour. See also Bleaching agents, Bread improvers and Dough conditioners.

Metal scavengers: See Sequestrants.

Mineral supplements: See Nutrient supplements.

Moisture-retaining agents: See Humectants.

Mold inhibitors: See Antimold agents.

Mold retarders: See Antimold agents.

Natural colors: See Colors, food.

Neutralizers: Adjustment of acidity is necessary in the production of some dairy products. For example, excessive acidity that may develop in cream must be neutralized for satisfactory churning to produce a butter of acceptable flavor and keeping quality. Neutralizers are also called neutralizing agents. See also Acidulants and Alkalies.

Neutralizing agents: See Neutralizers.

Nonnutritive Sweeteners: Sugar substitutes are permitted in foods to add sweetness but neither calories nor other food values. At one time, nonnutritive sweeteners and foods containing them were limited to persons who were obliged to restrict their intake of ordinary sweets. Due to governmental laxity, many foodstuffs with nonnutritive sweeteners came into widespread use by the general public. Nonnutritive sweeteners are added to certain beverages, canned fruits and vegetables, frozen desserts, gelatin, jams, jellies, marmalades, baked goods, salad dressings and frozen fruits.

Soft drinks sweetened with nonnutritive sweeteners do not have the same mouth feel as those sweetened with sugar. Some processors adjust this by adding bodying agents, while others raise the carbonation to an unusually high level.

Nonnutritive sweeteners have a poor safety record. Some, long in use, were ultimately banned because of their harmfulness. Saccharin is currently under review (*see* pages 91-97). Presently, more than a dozen new nonnutritive sweeteners are being studied in laboratories. Nonnutritive sweeteners are also called artificial sweeteners, sugar substitutes, and synthetic sweeteners.

Nutrient supplements: Vitamins and minerals are added to some foods to replace those lost in processing. The addition began on a large scale in the early 1940s, when a number of companies initiated the "enrichment" program with white flour, degerminated cornmeal, white rice and other grains that had been impoverished through over-refinement.

Ever since the synthetic synthesis of vitamin D, dairies have engaged in a program of "fortification" of fluid and evaporated milk. Not to be outdone, other food processors began to fortify breakfast cereals, bread, macaroni, margarine, ice cream, prepared vegetables for infants, and at times, even hot dogs. Warning was made concerning the dangers of excessive intake of vitamin D, especially by infants. But it took several decades before the FDA took action. At last, the agency requested a review by joint committees of the American Medical Association (AMA) and the American Academy of Pediatrics. Upon the recommendation of these two groups, the FDA ordered a reduction of vitamin D fortification in foodstuffs. But in July 1968, the FDA *withdrew* its proposal to remove vitamin D from the GRAS list, although it *should* have been removed.

Oleoresins: Instead of using spices for flavorings, sometimes their oleoresins are used. This is done by percolating a volatile solvent through a ground spice. The solvent is vacuum-distilled off, and what remains is the oleoresin. Although the oleoresin is very similar to the spice from which it is derived, it is not identical. Not all of the spice's flavors are extracted. Certain spices are extracted to yield their oleoresins, not for the flavoring effect, but for the intense colors produced. Examples are paprika and turmeric, used in French-type salad dressings. See also Flavorings, food.

Oxygen interceptors: See Antioxidants.

pH adjusting agents: See Alkalies.

Plasticizers: Substances are added to food to impart flexibility, workability or distensibility. Plasticizers are used in edible coatings for meat and cheese; in tablets used for food; and in chewing gum. They are used in pet food to retain a desired texture over a long period of time. Plasticizers are also called softeners.

Polishing agents: Substances are used to form a glossy, hard finish, especially to coat candy. These agents are also called glazing agents. See also Coating agents.

Preservatives: Substances are added to foods to delay or prevent undesirable changes in them. A number of different types of preservatives are used, depending on the food product and the spoilage organism involved. Some preservatives delay food spoilage caused by the microbial growth of molds, yeasts, rope and bacteria (as in cheeses, fruits and baked goods) (*see* pages 41-43 and 52-53). See Antibiotics, Antibrowning agents and Antimold agents. Other preservatives retard oxidative rancidity in food (as in oils and fats) (*see* pages 43-44 and 44-46). See Antioxidants. Some preservatives prevent certain physical or chemical changes that affect color, flavor, texture or appearance (as in meats, soft drinks and dried fruits) (*see* pages, 102-106, 41-44 and 108-112). See Fumigants and Sequestrants. Other preservatives prevent the growth of fungus spores (as on citrus fruit rinds). See Fungicides. Time-honored preservatives include salt, sugar, vinegar, spices and wood smoke. However, long-time use is no assurance of safety. Wood smoke, for instance, contains carcinogens.

Propellants: Gases are used in pressurized containers of whipped cream, dairy and vegetable-fat toppings and other foamed or sprayed food products. Propellants are also called pressure dispensing agents. See also Aerating agents.

Release agents: Oils, petroleum derivatives and dry powders are used on hard-candy products, other confections, and on baking pans to prevent products from sticking. Release agents are also called antisticking agents, dusting agents and lubricating agents. See also Coating agents.

Rope inhibitors: See Antirope agents.

Sequestering agents: See Sequestrants.

Sequestrants: Metallic ions, such as iron and copper, are present in certain food products. If they are allowed to remain, even in low concentrations, these metals can markedly reduce the shelf life of fats, oils and other foods that oxidize and spoil. Sequestrants are substances that tie up and deactivate these metallic impurities, and also improve the flavor, color and texture of foods. However, by tying up the metals, sequestrants may interfere with the absorption of certain trace elements important in good nutrition (*see* pages 50-52). Sequestrants are also known as sequestering agents, chelators, chelating agents and metal scavengers.

Softeners: See Plasticizers.

Solvents: Certain substances are capable of dissolving or dispersing one or more substances. Water is a solvent for many salts; alcohol, for many resins; and ether, for fats. Solvents are used to extract substances from flavorings and spice oleoresins, colors, vegetable oils and other foods. In general, solvents are toxic materials in large amounts, and undesirable as residues in foods and beverages.

Stabilizers: Traditionally, stabilizers have been used to keep certain mixtures from separating. Stabilizers prevent the cocoa from settling to the bottom of chocolate milk drinks. The orange pulp is kept from settling to the bottom of orange drinks.

Synthetic, nonnutritive stabilizers now reduce food processing costs by replacing more expensive traditional ingredients. They help bind the water in ice cream and prevent the formation of large, grainy ice crystals. In this manner, stabilizers added to ice cream help produce a smooth creamy texture in the product that is usually associated with expensive ingredients such as cream and milk.

Some of the traditional natural stabilizers as well as the newer synthetic ones are now under official review (*see* pages 53-58, and 100-102). Stabilizers are also called stabilizing agents and suspending agents. They are frequently used with emulsifiers. See also Sequestrants.

Stabilizing agents: See Stabilizers.

Sugar substitutes: See Nonnutritive sweeteners.

Supplements, nutrient: See Nutrient supplements.

Surface active agents: See Emulsifiers.

Surfactants: In some instances, surfactants decrease foam; in other cases, they create foam. See Emulsifiers. See also Foaming agents and Defoaming agents.

Suspending agents: See Stabilizers.

Synthetic colors: See Colors, food.

Synthetic sweeteners: See Nonnutritive sweeteners.

Texturizers: Casein, the principal protein of cow's milk, is used to give texture to ice cream, frozen custards, ice milk and fruit sherbets.

Thickeners: See Emulsifiers, Stabilizers and Vegetable gums. Thickeners are also called thickening agents and bodying agents.

Thickening agents: See Thickeners.

Vegetable gums: Certain plants, from both sea and land,

yield substances used as thickeners, stabilizers and water retainers in a wide variety of processed foods, including beverages, frozen desserts, puddings, meringues, icings, baked goods, frozen candied sweet potatoes, cheeses, cheese spreads, cheese foods, salad dressings, confections, jellies and chewing gum. See also Stabilizers.

Viscosity agents: See Emulsifiers.

Vitamin supplements: See Nutrient supplements.

Washing agents: Chemical compounds that act as preservatives are used to wash fruits and vegetables, and to wash the curd in cottage cheese to prepare these foods for processing. In concentrated form, the washing agents are toxic. Although they are used in diluted form, there is no assurance that residues do not remain. Nor is there guarantee that such residues, ingested repeatedly, are innocuous.

Water-retaining agents: See Humectants.

Wetting agents: See Emulsifiers.

Whipping agents: See Aerating agents.

Yeast foods: Chemicals are used by both brewers and bakers to manipulate the fermentation process for their own convenience. Yeast foods have no counterpart in home baking or traditional brewing, and merely increase the body burden of unnecessary additives.

A Glossary of Medical Terms

adipose: the fat in the cells of human and animal tissue

ataxia: the failure of muscular coordination

carcinogen: any cancer-producing substance

carcinogenesis: the production of cancer

carcinogenic: producing cancer

carcinogenicity: the state of cancer

carcinoma: a malignant tumor, or cancer

cecum: a pouch at the beginning of the large intestine

cocarcinogen: a substance that is normally not a cancer-inciter, but, in combination with other substances, may become a cancer-inciter; also called a weak carcinogen

collagen: the main organic constituent of connective tissue and of organic substance of the bones

DNA: deoxyribonucleic acid, substances associated with the transmission of genetic information

edema: the presence of abnormally large amounts of fluid in the body; swelling

EFA: essential fatty acids, substances essential both for the physical structures of the body and for many of its biochemical processes

fibrosis: the formation of fibrous tissue

granuloma: a tumor made up of granulation tissue

Heinz bodies: small bodies occasionally seen in red blood corpuscles, sometimes associated with anemia

hyperplasia: the abnormal multiplication or increase in the number of normal cells in normal arrangement in a tissue

metabolism: the sum of all the physical and chemical processes by which living organized substance is produced and maintained; also the transformation by which energy is made available for the uses of the organism

mutagen: an agent that tends to increase the occurrence or extent of mutation

mutagenesis: the occurrence or induction of mutation

mutagenic: capable of producing mutation

mutagenicity: the state of producing mutation

mutation: a sudden and fundamental change in the genes

necrosis: death of a circumscribed portion of tissue

papilloma: a benign tumor, such as a wart, resulting from an overgrowth of epithelial tissue

peritoneum: the membrane that lines the abdominal walls

polyp: a smooth and pedunculated growth from a mucous surface

sarcoma: a tumor, often highly malignant (cancerous)

tachycardia: excessive rapidity in the heart action

teratogenesis: the production of monstrous growths or fetuses; the production of birth defects

teratogenic (or teratogenetic): relating to teratogenesis; tending to cause monstrosity (defective offspring; birth defective infants)

teratological: relating to the abnormality of organic growth or structure

tonic-clonic (convulsions): convulsions marked by contractions-relaxations

urticaria: hives

weak carcinogen: see cocarcinogen

Identification of Acronyms

AMA	American Medical Association
BHA	butylated hydroxyanisole
BHT	butylated hydroxytoluene
CFDD	Canadian Food and Drug Directorate
CRS	Chinese Restaurant Syndrome
DNA	deoxyribonucleic acid
EFA	essential fatty acids
FAO/WHO	Food and Agricultural Organization/World Health Organization
FASEB	Federation of American Societies for Experimental Biology
FDA	Food and Drug Administration
FD&C	Food, Drug and Cosmetic (Color)
FEMA	Flavoring Extract Manufacturers Association
GRAS	Generally Recognized as Safe
IARC/WHO	International Agency for Research on Cancer/World Health Organization
MIT	Massachusetts Institute of Technology
MSG	monosodium glutamate
NAS	National Academy of Sciences
NAS–NRC	National Academy of Sciences–National Research Council

NTA	nitrilotriacetic acid
OTA	Office of Technology Assessment
ppb	parts per billion
ppm	parts per million
SCOGS	Select Committee on GRAS Substances
USDA	United States Department of Agriculture
USPHS	United States Public Health Service

Some Additives Commonly Listed on Food Labels

Of some three to four thousand intentional additives currently used in foods, relatively few are listed on labels. Federal regulations are irrational and chaotic. For example, if monosodium glutamate is used in canned vegetables it *must* appear on the label. However, the same additive may be used in mayonnaise, French dressing and salad dressing *without* label declaration. Many newly innovated convenience foods which are nonstandardized items will list the additives contained in them. Many standardized foods (bread and baked goods; jellies and jams; etc.) will not have the additives listed.

The purpose in listing the following additives commonly found on food labels is to alert consumers to the functional uses of these substances, and to known facts about their properties. What has been stated earlier bears repetition: the paucity of information about some additives results from two factors. First, toxicological data may be lacking. Second, even if data exists, the public has had no access to them from the FDA files.

Benzoic acid, a chemical preservative considered GRAS, is permitted up to 0.1 percent in margarine to retard flavor reversion; in fibered codfish, salt codfish, and the ice for cooling fish; in maraschino cherries; in mincemeat; in fruit juices; in pickles; in confections; in fruit jelly, preserves and jams in a quantity reasonably necessary as a preservative; in bottled soft drinks up to 0.05 percent; and in dietary supplements up to 0.1 percent. Benzoic acid is also considered GRAS as a flavoring in beverages; in ice cream and ices; in candy; in baked goods and icings; and in chewing gums. Benzoic acid is considered GRAS as a post-harvest fungicide and as an antimold in food packaging materials.

Benzoic acid occurs in nature in free and combined forms. Most berries contain appreciable amounts (about 0.05 percent). After ingestion, benzoic acid combines with glycine and is excreted mainly as hippuric acid by most vertebrates, except fowl. Human toxicity: a mild irritant to the skin, eyes and mucous membranes.

Scientists investigated the toxicity of benzoic acid in male weanling rats. Benzoic acid, fed at levels of 3 percent in the diet for five days, resulted in growth retardation, brain damage and neurological disorders shown by irritation, ataxia and tonic-clonic convulsions. Growth retardation and brain damage could still be seen nineteen to thirty days after treatment was stopped. At levels of 1.1 percent of benzoic acid in the diet of male weanling rats for thirty-five days, results were growth retardation and impaired food utilization, but no neurotoxic signs or pathological brain damage.

In Soviet tests, benzoic acid was fed to mice for eight months, and to rats for periods of three months and eighteen months. Benzoic acid was markedly toxic to the mice. Body weights were reduced and survival rate was below normal. Stress conditions were observed, and benzoic acid gave some indication of being a cocarcinogen. On the basis of these findings, the researchers recommended that the use of benzoic acid be restricted.

In later Soviet tests, short- and long-term feeding with benzoic acid added to the diet of mice and rats showed similar adverse effects, as in the previous study. Growth and survival rates were impaired, and the animals showed signs of stress. When benzoic acid was combined with sodium bisulfite, another food preservative, the effect appeared to be synergistic or additive: the toxic effect was generally greater than when each preservative was given singly. On the basis of these findings, the researchers recommended further restrictions in the USSR on the use of benzoates and sulfites as preservatives, and the substitution of innocuous ones. The recommendation was approved by the USSR's Food Hygiene Department of the Ministry of Health.

Butylated hydroxyanisole, also called BHA, is the most widely used antioxidant in foodstuffs in the United States. BHA is considered GRAS in specified amounts in beverages; ice cream; ices; with fat in candy; baked goods; gelatin desserts; potatoes; mixed and diced glacéed fruits; potato flakes and granules; sweet potato flakes; chewing gum base; defoamer component in yeast and beet-sugar production; dry breakfast cereals; dry yeast; dry mixes for beverages and desserts as well as beverages and desserts prepared from dry mixes; in emulsion stabilizers for shortenings and lard; unsmoked dry sausage, frozen fresh pork sausage and freeze-dried meats; nuts; soup mixes; food packaging materials, including plastic films, wax coatings, rubber gaskets that seal food jars, wax paper and as a component of lubricants with incidental food contacts; sometimes used in combination with other antioxidants (*see* Butylated Hydroxytoluene, pages 44-46).

Various experiments with animals have demonstrated the damaging effects of butylated antioxidants. BHA retarded the growth of weanling albino female rats and caused weight loss in adult animals. BHA was found to inhibit the contraction of smooth muscles of the intestine in the ileum area.

In reviewing the GRAS list, FASEB concluded in 1978 that while there is "no evidence to demonstrate that BHA is a hazard to the public when used at current levels," additional studies should be required for BHA's effects on the human enzyme and metabolic systems. FASEB regarded the safety questions about BHA as being "less urgent" than those about BHT (*see* Butylated hydroxytoluene).

Butylated hydroxytoluene, also called BHT, is a widely used antioxidant. BHT is considered GRAS used in specified amounts in chewing gum base; potato flakes and granules; sweet potato flakes; as defoamer component in yeast and beet sugar production; in dry breakfast cereals; in emulsion stabilizers for shortenings; enriched rice; animal fats and shortenings containing animal fats, frozen fresh pork sausage and freeze-dried meats; sometimes used in combination with other antioxidants (*see* Butylated hydroxyanisole, pages 43-44).

Various experiments with animals have demonstrated the damaging effects of butylated antioxidants. BHT seems to be even more toxic than BHA. Among many adverse findings were metabolic stress, depression of growth rate, loss of weight, increase of liver weight, damage to the liver and kidneys, increase of serum cholesterol and phospholipid levels, baldness and fetal abnormalities in offspring such as failure to develop normal eyes, and the storage of BHT in tissues, fat and organs with slow excretion of it.

Experimental tests in Romania revealed such metabolic stress in rats fed BHT in their diets that recommendation was made to ban the antioxidant from food use in Romania. It is not permitted in Sweden or Australia. In the United Kingdom the Food Standards Committee recommended a ban on all further uses of BHT in foods, but an outright ban was never implemented. Rather, the use of BHT was restricted to limited use in butter, lard, margarine and essential oils. Several appeals from industry to extend its use to other foodstuffs were denied. In 1958, the British committee reexamined and reevaluated the data, and once again recommended the withdrawal of BHT from the list of permitted antioxidants. This recommendation was followed by a customary "full consideration to any representations made by the interests concerned," just as American industry is afforded an opportunity to protest a proposed regulation. British industry protested and the committee's recommendation was not acted upon. In 1963, the committee reviewed the subject again, and once more recommended that BHT should no longer be permitted because "the margin of safety for BHT is less than for the other permitted antioxidants." Two years elapsed. In 1965, the committee examined new evidence, and studied a recommendation to ban BHT entirely from food. However, because of the inconclusiveness of the evidence, the committee withheld final judgment until ongoing experiments are completed. Although no outright ban was made, BHT was banned from baby foods in Great Britain, and its amount in other foodstuffs was reduced by half.

In 1972, scientists reported that when pregnant mice are fed diets containing large quantities of BHA and BHT, their offspring frequently suffer major changes in brain chemistry and show abnormal behavior patterns. The researchers concluded that "these antioxidants can no longer be viewed as innocuous substances."

Regarding human toxicity of BHT, it "may cause sensitization-type dermatitis, practically no systemic toxicity." But severe allergic reactions have been reported for BHT (and BHA), including "debilitating and disabling chronic asthmatic attacks," skin blistering, eye hemorrhaging, tingling sensations on face and hands, extreme weakness, fatigue, edema, chest tightness and difficulty in breathing.

As a result of FASEB's review in 1977 the FDA removed BHT from the GRAS list. The FDA proposed to restrict use of this antioxidant to current levels in foods for which it had already been approved, pending the completion of newly-required safety studies by manufacturers. The new studies were intended to resolve such questions as whether BHT causes changes in the human liver. This effect had been found in rats fed BHT, but rats do not break down the chemical in the body the same way that humans do. The new studies would attempt to learn which species of test animals are appropriate for studies of how BHT reacts in humans, and then tests using these animals would be required. Under the FDA's proposal, any companies wishing to use BHT must perform the needed studies; otherwise, the FDA would seek to ban BHT as a food additive.

Since the FDA gave BHT interim status—to continue in use while tests were being conducted—public interest groups objected and petitioned for an immediate ban. The FDA action was termed "halfhearted" in view of damaging data already available regarding BHT's lack of safety. There was some evidence that the antioxidant causes tumors in test animals. Furthermore, BHT oxidizes to form a chemical (3,3,5,5 tetra bisstilbene-quinone) that has not been tested. The public interest petition pointed out that other substances and techniques existed to extend the shelf life of fats and oils, which would be safer. These included use of citric acid, vitamin E, use of dark bottles, packing under inert gas such as nitrogen, and use of refrigeration.

Caffeine, a stimulant, is considered GRAS for use in certain nonalcoholic, carbonated beverages (soft drinks); and in cola-type beverages up to 0.02 percent. Caffeine occurs naturally in tea, coffee and maté leaves as well as in cola nuts.

Under the Standards of Identity for Nonalcoholic Carbonated Beverages, caffeine may be added as an optional ingredient. Its presence *must* be stated on the label. However, the cola-drink manufacturers successfully pressured the FDA to allow them the special privilege of avoiding this label declaration. Thus, although it is required by Federal law that cola drinks *must* contain caffeine, its label declaration is *not* mandatory. Although there is some caffeine present naturally in the cola nuts, additional caffeine may be put into the cola drinks. Their caffeine content has been found to be as much as 45 to 50 milligrams in an eight-ounce glass, compared to 100 milligrams in a cup of coffee and 110 milligrams in a typical "stay awake" tablet. Uninformed parents, who may prohibit their children from drinking caffeine-containing coffee or tea, may allow them to consume caffeine-containing cola drinks.

In high concentrations, caffeine was shown to be mutagenic for *Escherichia coli* (a microorganism), and was shown to inhibit its DNA repair. In test-tube studies with cells from mouse and from man, DNA repair was shown to be inhibited. But it is difficult to extrapolate data from nonmammalian systems to mammalian ones. In feeding tests with mice, caffeine failed to show mutagenic effects.

However, caffeine did show signs of teratogenicity in mice, dependent on the route of administration. (Substances may be administered by oral ingestion, skin painting, subcutaneous injection, pellet implantation, etc.) Since the routes of administration can have different effects, results are inconclusive.

Caffeine has produced gastric ulcers when fed experimentally to cats, guinea pigs and rats.

The human toxicity of excessive caffeine is well known. As side effects, caffeine may produce nervousness, insomnia and tachycardia. Caffeine frees fatty acids from adipose tissues and causes a rise of fatty acids in the blood. This action results in fatty deposits in the liver. In a review of caffeine research for a soft-drink company, a professor of pharmacology was quoted as saying, "Although caffeinated soft drinks provide a lift that refreshes and fights fatigue fairly well, they produce a tremor, insomnia, gastrointestinal disorder and . . . may play a role in death from cancer or heart stoppage."

By law, caffeine must be added to cola drinks. Therefore it is regulated as a food additive for this purpose. Since caffeine is a naturally-occurring constituent in coffee, tea, and chocolate, the caffeine in these beverages is not subject to food additive regulations.

In June 1978, SCOGS submitted its final report to the FDA on caffeine in cola drinks. Citing a dearth of research on any long-term effects of caffeine consumption, and possible behavioral effects on children, SCOGS recommended that caffeine be withdrawn from the GRAS list. A major concern was that *the highest consumption of cola-type beverages is by children from ages one to five years.*

The SCOGS group was more concerned about what is *not* known about caffeine consumption than what is known. The group viewed with concern the greatly increased consumption of caffeine in cola-type beverages, which represents "a unique addition to food of a pharmacologically active central nervous system stimulant." The amount of caffeine consumed in these beverages alone "borders on the dose known to produce central nervous system stimulation in animals and man. Whether such stimulation constitutes an adverse effect or whether a potential hazard may exist for the segment of the population, particularly children . . . exposed chronically to stimulating doses of caffeine, cannot be answered on the basis of the evidence now available."

Stating that the potential behavioral effects of caffeine consumption by children are neither adequately documented, nor their consequences clear, SCOGS pointed to the likelihood of possible effects for which more research is needed. The group recommended "rigorously controlled chronic fetal, neonatal (newborn), and growing-animal studies of appropriate species" in order to gather data on the immediate and ultimate behavioral and cardiovascular effects of caffeine added to the diet and given in cola-type beverages.

SCOGS' review of mutagenic and teratogenic studies failed to reveal any hard evidence. However, the group mentioned a recent, unpublished report indicating that rats given caffeine orally at daily doses of 150 to 250 mg/kg for fifteen months develop cancer of several organs. In long-term animal feeding studies, there was an increased embryonic loss in pullets fed caffeine, and a "marked but reversible reduction in sperm output and concentration in roosters fed caffeine."

SCOGS' concern with the behavioral effects of caffeine in cola beverages on children was greater than for adults, since "in children there can be chronic consumption of caffeine in cola-type beverages during the period of brain growth and development." During this period, the developing central nervous system is most sensitive to the effects of all aspects of the environment. *The estimated levels of caffeine intake at these ages are near those levels that are known to cause central nervous system effects in adults.*

Of the eleven scientists who served on the SCOGS review, one member, Ralph G.H. Siu, filed a minority opinion. Dr. Siu felt that by restricting the evaluation of caffeine merely to its addition in cola beverages, the SCOGS review was "an artificially bounded abstraction" with conclusions of "limited, even misleading, operational meaningfulness in real life." By excluding coffee, tea, chocolate, prescription medications, over-the-counter analgesics, cold preparations and stimulants, the effects of the total body burden of caffeine were not explored.

Beverages such as cola drinks account for only one-fourth of the total caffeine consumed from foodstuffs and covers less than one-fifth of the principal exposure time over an individual's lifespan.

Siu, in his minority opinion, presented some health problems *not* included in the SCOGS report. Caffeine affects many metabolic processes, including biochemical equilibrium. Caffeine reduces the blood-brain barrier permeability to an alkaloid, harmine, in rats, and sensitizes the mammalian respiratory center to carbon dioxide. Siu described many other effects of caffeine on the body, noting that singly such changes may appear innocuous. However, "in the face of potential synergistic concatenations," one or more of the effects "may initiate an unhealthy cascade." The short-term anti-fatiguing stimulation of caffeine may be harmless; but the long-term postponement of the natural need for rest and/or repair may lead to "physiological bankrupting" of the body. High coffee ingestion produces symptoms indistinguishable from anxiety neurosis. Many of the enhancing effects of caffeine on mutagenicity, chromosome damage, carcinogenicity, and teratogenicity of other agents, are apparent at levels of consumption only six to ten times that of heavy coffee drinkers. Hence, the usual hundredfold safety factor, estimated by toxicologists for substances that may inflict harm, is not being met. Siu recommended that official re-evaluation of the potential health hazard of caffeine in cola-type beverages alone is insufficient to determine the total adverse effects of caffeine on human health.

Calcium disodium EDTA, a sequestrant, is used to promote color, flavor, and color or texture retention in canned carbonated soft drinks; in distilled alcoholic beverages; in vinegar; in canned white potatoes; in cooked canned clams, crabmeat and shrimp; in cooked canned mushrooms; in cooked canned dried lima beans; in pecan-pie filling; in processed dry pinto beans; in spice extractives in soluble carriers; in pickled cucumbers and cabbage; in hard-boiled egg products; in artificially colored and lemon- or orange-flavored spreads; and in aqueous multivitamin preparations. Calcium disodium EDTA is used as an antigushing agent in beer; as a preservative in French dressing, salad dressing, mayonnaise and nonstandardized dressings; in margarine; in potato salad; in sandwich spreads and sauces. It is also used as a preservative in combination with other preservatives for salad dressings, sandwich spreads and sauces.

Calcium disodium EDTA fed to young rats produced liver lesions, kidney irritation, and degenerative lesions in the renal tubes of the kidneys.

Calcium disodium EDTA is used with humans as a detoxicant for lead and other heavy metals. This therapeutic use (especially the sodium salt) has produced well-documented cases of kidney damage. In excess, this substance can cause mineral imbalance, and produce a state similar to vitamin deficiency. It is a known enzyme inhibitor and a blood coagulant inhibitor. Known side effects include gastrointestinal disturbance, muscle cramps and kidney damage. Metabolic studies show that this substance is rapidly eliminated from the body. In oral doses given to man, the rat and the rabbit, 90 percent or more passed unabsorbed through the intestine. The remainder appeared in the urine, mostly within forty-eight hours after its ingestion.

The potential hazards of this food additive are created by many minute quantities, ingested in a variety of foods and beverages daily, over a prolonged period of time. In particular, the effect of such an intake must be considered for its effect on the absorption of essential trace elements such as iron, zinc, and copper. EDTA is so widely used in food processing that currently the average American may consume a hundred milligrams or more of this sequestrant daily. In addition, more than forty more sequestrants are officially approved for use by processors in foods and beverages. The extensive use of sequestrants in the food supply makes it difficult, or even impossible, to obtain adequate levels of trace minerals. Food refining is strongly suspected of creating states of deficiencies for certain trace minerals. This problem, with its grave health consequences, may be *compounded* by the extensive use of sequestrants such as EDTA.

Calcium propionate, a chemical preservative, is considered GRAS as mold and rope inhibitor in bread, rolls and other baked goods; in poultry stuffing; in chocolate products; in processed cheeses; in cake and cupcakes; in artificially sweetened fruit jelly and preserves; in pizza crust and in food packaging materials. Calcium propionate is sometimes used with sodium propionate.

Both the FDA and food processors are quick to affirm that calcium and sodium propionates are salts of propionic acid, which is produced naturally in cheese during its manufacture. This is stated to reassure the public that the additives are harmless *natural* substances. The propionic acids used by bakeries are made synthetically by large chemical and pharmaceutical companies, from ethylene, carbon monoxide and steam. Since fungi and molds are closely related, it is not surprising to learn that these materials are also common ingredients in solutions, powders and ointments used to treat athlete's foot and other mycoses.

A physician has reported allergic reactions from propionates, disturbances that began in the upper gastrointestinal tract, four to eighteen hours after eating foods containing propionates, and ended with partial or total migraine headaches. Since the gastrointestinal distress symptoms are similar to the gall-bladder attack, the symptoms may be especially severe in cases where an allergy to the propionate is combined with a gallbladder ailment. The physician charged that authorities failed to find harm in the propionates simply because they failed to investigate them.

Furthermore, when the sodium salt of propionic acid is used in food, the sodium places an additional burden on the body. When the calcium salt of propionic acid is used, it destroys the enzyme that normally makes it possible for the body to assimilate any calcium—present naturally or added through enrichment—in the bread.

These preservatives help cut baker's losses by reducing the return of stale baked products and other food items in which the propionates are used. As mold retarders, they can give stale goods the appearance of freshness.

Carrageenan, also known as Irish moss, has had a long history of use as a food in Western Europe and in the United States. Because of its jellying properties, this seaweed, when cooked with milk and seasoning, has become a popular custard, blancmange. Supplies of carrageenan were carried by the colonists to the New World and continued to be imported until some choice sources were discovered along the coasts of New England and the Maritime Provinces of Canada. Its harvesting began there as early as 1835.

Due to its mucilagenous quality, carrageenan also has had extensive medicinal use. It has been given for diarrhea, urinary disorders and chronic chest inflammations, including tuberculosis.

In recent times carrageenan and its extracts have come into widespread use as a food additive, for these substances have characteristics that serve many different needs in food processing. Carrageenan will form a gel above room temperature, and it will not become soft or melt after it is unmolded and placed on the table. It can be packed under high-temperature sterile conditions and stored for long periods (as dessert gels for babies). It can hold high amounts of water. It mixes well with sugars, starches and other carbohydrates; vegetable gums; milk, eggs, gelatin and water-soluble salts.

As with many other food additives, carrageenan was *assumed* to be innocuous because of its plant origin, its long-time use, and its *apparent* record of safety. Toxicity studies for it were begun only as late as 1959. High levels (25 percent) of carrageenan in the diet of mice and rats produced liver lesions in a few animals, which suggested cirrhosis. Most of the animals showed a retarded growth rate. But since carrageenan was known to depress peptic activity, investigators attributed the adverse effects to dietary deficiency. Although some animals died, there was no consistent cause of death. Nor were colonic ulceration or scarring produced.

Ten years later, another experiment with carrageenan produced totally different results from the earlier studies. The new findings showed great discrepancies of results between the two studies. Concentrations of only up to 5 percent degraded carrageenan in the drinking water caused weight loss within two to three weeks in guinea pigs and rabbits, but not in rats or mice. However, all four species developed pinhead-sized ulcers in the cecum, the colon and the rectum. These lesions resembled those found in human ulcerative colitis. At later stages, other pathological changes developed in some animals. These findings were considered important in view of an increasing incidence of ulcerative colitis in man.

The same investigators made further studies. One group of guinea pigs was given low levels (1 percent) of undegraded carrageenan in their drinking water while another group was given higher levels (5 percent) of degraded carrageenan in theirs. Both groups developed ulcerative lesions, beginning in the cecum and spreading to the lower colon and rectum. Although such a distribution is not the typical pattern in human ulcerative colitis, the tissue damage in the animals, when examined microscopically, appeared similar to that in man. Other effects on the guinea pigs included blood and mucous in the feces, and retarded growth.

Then the researchers gave rabbits carrageenan in drinking water. A low level (1 percent) produced diarrhea and ulceration in all the rabbits. Even at one-tenth that level (one-tenth of 1 percent), ulceration was produced in half the rabbits. Lesions were mainly in the cecum. Microscopic examination showed an infiltration of carrageenan in the cells, and the presence of abscesses and polyps.

These findings created a controversy that continues to rage. Other studies have produced conflicting evidence. Some experiments have reaffirmed earlier findings of the antipeptic activity of carrageenan, but have failed to confirm ulceration.

Because carrageenan is known to inhibit pepsin and other gastric enzymes in the human stomach, it has been used therapeutically to treat gastric ulcers. Proponents say that this therapy has been used with many patients who do not appear to have suffered ill effects. To reaffirm this argument, degraded carrageenan was fed experimentally to baboons at levels used therapeutically for humans. There was little absorption from the gastrointestinal tract.

Despite its use with gastric ulcers, carrageenan is known to have inflammatory effects. Injected under the skin, it readily produces an inflammation. This suggests that *in high concentrations it has an irritant effect when it is in direct contact with tissues*. Tests with animals have confirmed this. Carrageenan, injected in rats directly under the capsule of the liver, led to the formation of liver fibrosis. The connective tissue formed with carrageenan was reabsorbed after the injected material disappeared.

In other studies, carrageenan produced tumors. In guinea pigs, a single injection of carrageenan under the skin led to the formation of collagen granuloma. In rats, sarcomas developed at the site where carrageenan was injected.

Further tests are scheduled by carrageenan manufacturers. In the wake of the cyclamate ban in 1970, the FDA requested a review by the National Academy of Sciences–National Research Council (NAS–NRC) of many food additives, including carrageenan, about which doubts have been raised.

To the present, what is definitely established is that ulcers *can* be induced by carrageenan under certain conditions in some animals. Animals apparently possess an ability to adsorb this material in their intestinal tract. Whether this holds true for man remains to be demonstrated. Meanwhile, consumers are being exposed to a possible hazard of unknown dimension. Ulcerative colitis is a particularly unpleasant, serious, and sometimes fatal condition. This question has been raised but remains unresolved: is the current widespread use of carrageenan, and its related substances, as a food additive contributing to human ulcerative colitis?

Carrageenan was among the first GRAS list substances to be reviewed following the presidential mandate of 1969. Carrageenan was tested for the FDA by the Food and Drug Research Laboratories, Inc., and results indicated that birth defects were induced when carrageenan was fed to various species of mammals at high dosages. However, using the traditional hundredfold safety factor to extrapolate animal results to humans, investigators concluded that present levels of use presented no human health hazard. The birth defect finding forced the FDA to remove carrageenan from the GRAS list in June 1972.

In another study, conducted by the Institute of Experimental Pathology and Toxicology at Albany Medical

College, it was found that highly degraded, chemically altered carrageenan caused tissue changes in rats.

Carrageenan became suspect as a health hazard when the FDA's own tests produced liver lesions in laboratory rats. This finding caused a consumer magazine to urge its readers not to eat foods containing carrageenan as an ingredient until its safety has been proved.

In 1978, in a General Accounting Office report, carrageenan as well as the calcium salts of carrageenan were listed as suspected carcinogens by the International Agency for Research on Cancer, WHO (IARC), based on animal studies.

Carrageenan, a versatile food additive, is used as a suspending agent to keep chocolate from settling in chocolate milk and chocolate-flavored milk drinks; to stabilize fruit juices, fruit nectars and soft drinks. It is used as a clarifier in beet juice, vinegar, wine and beer, as well as to stabilize and control beer foam. It is used as a crystal inhibitor in frozen desserts such as ice cream, frozen custard, sherbet and ice milk. It stabilizes milk protein and prevents whey separation. It is used as a stabilizer to prevent oil-water separation, to minimize surface hardening, and to improve the spreading properties in cream cheese, other soft cheeses (such as neufchâtel type), processed cheeses, cheese spreads and cheese foods; to induce small-curd formation in cottage cheese; to prevent contrasting streaks of colors or flavors from mixing in variegated ice creams, syrups and purees. It is used as an emulsifier and stabilizer in pressure-dispensed whipped cream, yogurt, evaporated milk, coffee whiteners, filled and imitation milk and whipped toppings. It is used as a gelling agent in water-, milk- and starch-based puddings. It is used as a bodying agent to improve mouth feel in soups, sauces, gravies, soft drinks, syrups, and toppings; to thicken salad dressings, relishes, mustard and catsup; to improve mouth feel in low-calorie foods that lack sugar and oil; to replace all, or part of the starch in low-calorie puddings and pie fillings; to provide bulk for a feeling of satiety and to mask the aftertaste of artificial sweeteners in low-calorie foods. It is used as a water binder, adhesive and emulsifier to improve the appearance of processed meats; a gel in fillers and binders for sausages, caviar and pet foods; to prevent textural breakdown in canned fish, poultry, soft meats (such as tongue) and canned solid, ground and jellied meats. It is used as an adjunct to antioxidants in edible coatings on frozen poultry and meat; with antibiotics to improve their uniform distribution on fish. It is used as a flour conditioner that acts on the flour protein during baking to improve the dough strength, loaf volume, shape and finished texture in a variety of baked goods. It is used as a texturizer for cake batters; to reduce the fat absorption and improve the texture of yeast-raised doughnuts; to give a moist texture and even distribution of fruits in fruit cakes; an adjunct to other bakery ingredients (such as fruit and pie fillings, bakery jellies, toppings, icings and citrus oil emulsions). It is used as a sugar-crystal inhibitor in certain confections (with caramels, nougats and taffys); to emulsify fat and keep it evenly distributed throughout certain confections; also in jelly candies, candy bars and marshmallows. Carrageenan, in addition to its food uses, is also of great importance to the cosmetic and pharmaceutical industries. It is used in many consumer goods, including toothpastes, hand lotions, deodorants, spermatocidal jellies and medications.

Certified Colors: The words "U.S. Certified Color" on a food label *should* be a reassuring phrase to consumers regarding the safety of a food color. Unfortunately, the history of U.S. Certified Food Colors has been, and continues to be, a story of public consumption of toxic substances.

The chemical classes of coal tar dyes, from which the U.S. Certified Food Colors are made, are azo, triphenylmethane and xanthene. All three classes have shown potentialities for harm.

It has long been known that certain azo colorings can induce liver cancer in rats. The coloring is bound to the protein of the liver. The toxicity of the azo dyes depends on how they are metabolized within the body. Some merely pass through, and are excreted in an unchanged form. Others break down into innocuous products, and are excreted. Some are reduced by flora in the gastrointestinal tract. But at times azo dyes are absorbed into and bound to organs and tissues of the body. This is the action that can do harm. A number of azo colorings produce foreign particles in the blood, known as Heinz bodies. Some have produced sarcomas at the site of repeated injection of the dye under the skin of experimental animals. There is some disagreement among experts as to whether the production of sarcomas by this technique necessarily indicates that the substance is cancer-inciting, but, combined with other findings, it certainly should serve as a warning.

On the basis of animal experiments, triphenylmethane coal tar colors are potential human carcinogens. Xanthene coal tar colors have demonstrated mutagenic properties.

In 1900 some eighty coal tar dyes were used in the United States for food coloring. No regulations existed regarding the nature or purity of the dyes. The same batch of color used to dye cloth might find its way into candy.

The first comprehensive legislation, in 1906, reduced this list to seven dyes permitted for food use. These were chosen after study and testing which, naturally, was limited to the procedures and knowledge current at that time. Of the seven dyes originally approved, four, at later dates have been shown to be harmful, and have been prohibited from further food use.

From 1916 to 1929, ten more coal tar dyes were added to the original list. Of these, five, at later dates have been shown to be harmful, and have been prohibited from further food use.

Although the coal tar food colors continued to be widely used, officials acknowledged lack of information regarding their metabolism, excretion and toxicity. A United Kingdom report on food colors in 1954 warned that "we cannot accept the contention that, because coal tar colors have been used in foods for many years without giving rise to complaint of illness, they are, therefore, harmless substances. Such negative evidence in our view merely illustrates that in the amounts customarily used in foods the colors are not acutely toxic but gives no certain indication of any possible chronic effects. Any chronic effects would be insidious and it would be difficult if not impossible to attribute them with certainty to the consumption of food containing coloring matter."

It became apparent, in fact, that a large number of coal tar colors *were* toxic. In the 1950s, the FDA initiated an animal testing program that required more testing and higher feeding levels than the earlier pharmacological studies had used. The agency redefined "harmless" to mean a substance incapable of producing harm in test animals *in any quantity* and *under any conditions*. As a result of these new higher standards, eight coal tar dyes were delisted for food use. It soon became apparent that a continued full implementation of the standard would inevitably lead to the delisting of most, if not all, coal tar food colors. The Certified Color Industry Committee, comprised of producers who manufacture over 90 percent of all United States certified food colors, worked closely with the FDA to "modify" the testing standards, and to establish "safe" limits for the use of certified food colors. Jointly, the two groups formulated legislation. When enacted, this became the Color Additives Amendment in 1960, Public Law 86-618.

The new legislation weakened consumer protection by providing for less stringent interpretation of "harmlessness." Hereafter, human safety would be determined by a safety factor established by a "no-adverse-effects-level" in animals. Furthermore, food colors in use would be granted "provisional" listings. Under this practice, the FDA has allowed the continued use of food colors to be ingested by the public while tests are still in progress to establish their safety. Repeatedly, the FDA has granted liberal extensions of time for the completion of these tests. As a result of such extensions, some food colors have been in use "provisionally" for as long as nine years.

The safety record of currently used U.S. certified food colors is far from reassuring. The few remaining dyes, permitted in use on an interim basis, may ultimately join their predecessors as banned dyes.

—*FD&C Red No. 3* (*Erythrosine*) was provisionally approved by the FDA. Although the FAO/WHO Committee found this xanthene dye "acceptable for use in foods," the dye was granted only temporary acceptable status. The committee recommended further studies concerning the dye's metabolism with several species of animals, and preferably in humans. Additional studies were requested, on the mechanism responsible for the dye's effect on iodine levels bound to blood plasma. The German Research Institute for Food Chemistry in Munich found that erythrosine had a slight but significant mutagenic effect on a common microorganism, *Escherichia coli*. This finding is noteworthy since frequently, when substances are mutagenic, they are also carcinogenic. Elsewhere, at the National Institute of Neurological and

Communicative Disorders and Stroke, erythrosine was found to partially block the uptake of dopamine in brain cells of rats. This finding is consistent with the hypothesis that erythrosine can induce hyperactivity in some children.

—*FD&C Red No. 40 (Allura)* was permanently approved for food and drug use by the FDA as recently as 1971, and hailed by industry as "the only important new food color certified by FDA since 1939." In 1974, the FDA gave similar approval for this dye in cosmetics. The dye was developed to replace the banned FD&C Red No. 4, and became increasingly important when in 1976, FD&C Red No. 2 was also banned. Industry turned to FD&C Red. No. 40 as an alternate dye. Although this dye was given permanent clearance by the FDA, and approved in some other countries, the FAO/WHO Committee requested additional studies, and the Canadian government, which had examined the same data as the FDA, along with updatings, refused to approve use of the dye. The Canadians reported that the "evidence submitted by the manufacturer with respect to the safety of the product was inadequate." The Canadians were concerned especially with the long-term rat feeding tests designed primarily to determine if the dye caused cancer. Although the tests had been designed for a twenty-four month study, pulmonary disease ravaged the animals, and the tests had to be terminated. The Canadians questioned the adequacy of the tests, and voiced doubts that a twenty-one-month test was sufficient to detect a long-term effect such as cancer. One significant effect in these abbreviated tests was that the dye, fed at the highest level, caused moderate growth suppression during the first year. Although the FDA recommends full lifetime studies, conducted with two rodent species, the agency had granted permanent approval of this dye on the basis of studies with only one rodent species, the rat. The FDA, Canadian officials, and international food and health authorities requested further tests to provide additional data. Newer studies, termed "preliminary" by the FDA, showed that in week forty-one, of a seventy-eight-week study, a disproportionately high percentage of mice developed "premature and unexpected malignant lymphomas" (cancer of the lymph glands). Follow-up studies were inconclusive. FDA officials admitted that the dilemma with FD&C Red No. 40 bore a striking resemblance to the banned FD&C Red No. 2 (Amaranth), insofar as the safety of the dye had neither been proved nor disproved. A report, issued in February 1977 to the FDA, concluded that available data on mouse-feeding studies were insufficient to determine if the dye was carcinogenic. At Hazeltine Laboratories, a mouse study showed "borderline" significant increase in the number of carcinomas after seventy-six weeks. At a low level of FD&C Red No. 40, there was a significant increase of tumors in the hemopoietic (blood forming) systems of both male and female mice. Since results were equivocal, further studies were recommended before a final assessment.

The second interim report, made the following year, by an interagency group of scientists from the FDA and the National Cancer Institute (NCI), concluded that FD&C Red No. 40 does not cause cancer. In a Bioassay Program, conducted by NCI, para-cresidine, a chemical used to produce one of the intermediates that, in turn, is used to manufacture FD&C Red No. 40, was found to induce bladder cancer in rats and mice. However, the manufacturers of the dye reported that two stages in processing of the dye, with major chemical reactions, remove the offending chemical, which is not detected in the finished dye. Although the second interim report concluded that there is "no evidence at this time that

FD&C Red No. 40 is carcinogenic," the group hastened to add that prudence would demand that "a final assessment be made when the second mouse study is completed." This provisional clearance was greeted by one public interest group with the statement that the government panel is "playing a very deadly statistical game . . . and the odds are about 99 in 100 that this dye will be banned within the next year or two."

FD&C Citrus Red 2, an azo dye, was fed to rats and mice at different feeding levels for two years. At the highest level, retarded growth and reduced food intake were noted in males of both species. Serious bladder abnormalities were found in postmortem examination. An increased death rate for both sexes and an increased incidence of degenerative livers in females were noted. Repeated injections of the dye caused an increased incidence of early malignant tumors in females. As early as 1965, the FAO/WHO Committee classified this dye as one that had been "found to be harmful and that should not be used in food." In 1968, a British study showed that this dye tripled the rate of bladder cancer in rats. In the following year, the FAO/WHO issued a second, stronger statement, warning that since this dye "has been shown to have carcinogenic activity and the toxicological data available were inadequate to allow the determination of a safe limit, the Committee therefore recommends that it should not be used as a food color." Despite this recommendation, the FDA permits this dye to color orange skins not used or intended for further processing. The assumption is that the orange rinds will not be consumed, nor will the dye migrate into the fruit's edible portions. This assumption is faulty. Orange skins are used in marmalade, as candied rind, powdered rind, with roast duck, and in cocktails. At times, children peel oranges with their teeth, and inadvertently ingest the dye.

—*Orange B*, an azo dye, is a relatively new food dye, with research development begun in 1953. In 1966, the FDA approved its use in food, but restricted it to color the casings or surfaces of frankfurters and sausages at specified levels. Lifetime feeding studies of Orange B, with three animal species, failed to substantiate the safety of the dye. Liver nodules formed in dogs whose diets contained two percent or more of the dye. The significance of this nodule formation is unknown. The FDA stated, in the *Federal Register* on October 3, 1978, that the use of Orange B "could result in exposure of consumers to beta-naphthylamine. Based on information available at this time, free beta-naphthylamine (up to 2 ppm) is present in the intermediate used to produce the color." This substance is an acknowledged carcinogen. The FDA proposed to prohibit Orange B, and the manufacturer voluntarily stopped its production.

—*FD&C Yellow No. 6* (*Sunset Yellow FCF*) has been extensively studied for its effects on animals, but this azo dye has not been studied for metabolic effects in humans. In short-term studies with miniature pigs and with rats, as well as in acute toxicity studies with rats and mice, animals suffered from slight diarrhea, but no adverse effects were noted on growth or food consumption. In a seven-year study, dogs fed 2 percent of the dye in their diet suffered eye defects, which sometimes led to blindness.

—*FD&C Yellow No. 5* (*Tartrazine*), a pyrazolone dye, is one of the few dyes that has been universally accepted in countries where food coloring is permitted. In 1969, the FDA granted permanent listing for this dye. Nevertheless, tartrazine has been acknowledged as an allergen for a sizeable number of people. The FDA estimates that as many as 100,000 individuals in the United States may be allergic to tartrazine. For this reason, the agency proposed that by July 1, 1981, all food products containing this dye be required to list tartrazine specifically by name on the label; all drug products (except those applied only to the skin), by June 26, 1980. This decision is a landmark. It represents the first occasion on which the FDA has required that foods or drugs list a dye by name. Donald Kennedy, then Commissioner of Food and Drugs, explained on June 25, 1979, when the agency announced the regulatory action, that tartrazine "poses a particular hazard to some people" and the requirement "will enable those who are allergic to Yellow No. 5 to know which products contain it." Since May 1976, the FDA has required cosmetic products to identify tartrazine as well as all other dyes, by name, on labels.

Tartrazine may cause a variety of allergic reactions in persons who are sensitive to it, including urticaria, asthma, altered states of perception and behavior, uncontrollable hyperagitation and confusion. Frequently, persons who are sensitive to tartrazine are also sensitive to aspirin and benzoates.

Paul Khan, director of food protection, ITT Continental Baking Company, and self-styled "FDA Watcher" warned industry that the requirement to list tartrazine specifically on labels "may be the beginning of a broad push for color identification, with pressure for flavor and spice identification to follow."

—*FD&C Green No. 3 (Fast Green FCF)*, a triphenylmethane dye, produced a significant incidence of sarcomas at the site of repeated subcutaneous injections in rats. The FDA's chronic toxicity studies of this dye with rats, dogs, and mice showed it to be "without significant toxic effect" at the dosage levels fed, and the dye was listed provisionally.

—*FD&C Blue No. 1 (Brilliant Blue FCF)*, a triphenylmethane dye, induced a significant incidence of malignant tumors when injected at high levels in test rats. This dye is not permitted in the British Commonwealth, being considered "of probable toxicity." The dye was categorized as one "for which the data available are not wholly sufficient to meet the requirements" of acceptability for food use by the Joint FAO/WHO Expert Committee on Food Additives. FDA studies showed that the dye produced a high incidence of malignant tumors at the injection site, with some spreading mainly through the circulation system, and producing tumors in other areas. Despite a high mortality rate of the animals, the FDA granted provisional status. Later, the FAO/WHO Committee reviewed the dye and placed it in a category "found acceptable for use in food." In 1969, the FDA granted this dye permanent status.

—*FD&C Blue No. 2* (*Ingotine* or *Indigo Carmine*), is an indigo-related synthetic dye, provisionally accepted by the British Commonwealth, and placed by the FAO/WHO Committee in a category as a substance in need of "further information concerning its safety-in-use." The committee requested more data from a two-year study in a non-rodent mammalian species. Short-term studies showed that FD&C Blue No. 2 produced blood changes in pigs, and also liver abscesses. The FDA's two-year feeding studies with rats and dogs produced malignant tumors at the site of subcutaneous injection. Single oral doses of this dye were absorbed poorly by rats, with most excretion in the feces, and little in the urine or bile. It was difficult to identify breakdown products. But substantial quantities of the dye and smaller quantities of the breakdown production were excreted in the bile and urine when the dye was injected intravenously in rats. These findings caused the investigators to be quite concerned. Apart from the use of the dye with food, it has been employed in hospitals in the determination of kidney function. It had always been *assumed* that all injected dye was excreted, unchanged, by the kidneys. The presence of breakdown products in the bile and urine could distort results in kidney-function tests. Ingotine is provisionally listed for food use.

In late 1976, the FDA initiated a series of actions to produce a final determination on the status of some 80 dyes that for years have enjoyed provisional approval. Among these dyes, some included food dyes, while others were dyes used with drugs and cosmetics. When the actions are completed, ultimately, all provisionally approved dyes will either have been cleared for safety and granted permanent approval, or they will have been banned.

This action began on September 23, 1976, with the FDA's announcement that carbon black could no longer be used in foods, drugs, and cosmetics; nor the dye FD&C Red No. 4, in maraschino cherries or ingested drugs. Provisional approval was continued for a specified time for additional dyes, three of which would be permitted provisionally for food use.

This action was followed by the FDA setting a timetable, on February 4, 1977, for making final decisions on dyes. Life-time rat or mouse feeding and other studies would need to be completed for FD&C Green No. 3, Blue No. 2, and Yellow No. 6 by the deadline of January 31, 1981.

Industry saw the handwriting on the wall. Dr. Howard Mattson, public information director, Institute of Food Technologists, in addressing the institute's annual conference in June 1977, stated, "There's a possibility in six, eight, ten years . . . there won't be any synthetic dyes."

Forewarned, food and beverage manufacturers have displayed interest in reformulating some products by using colorants from natural sources. These include grape skins, cranberry pulp, dried beet root juice, carrot oil, cochineal extract, beta-carotene, annatto, curcuma extract, turmeric, and raisin juice.

A completely different approach is being developed by Dynapol, a high-technology research and development company in California, to produce food dyes and other food additives that are not absorbed during digestion in the human alimentary tract. A small additive molecule is attached to a much larger polymer molecule that cannot be broken into smaller molecules in the human system.

In 1972, Dynapol began working on food colors, antioxidants and sweeteners. The list of approved certified dyes has been dwindling rapidly due to banning of unsafe substances. As of 1979, only nine certified dyes remain, and the fate of these is uncertain. Unless alternatives are developed, many existing hues will disappear from food products. Dynapol's objective is to meet all or most of the spectral needs of food processors. Two colors have been developed by Dynapol using nonabsorbable polymers, a red-orange and a red-violet. At present, these polymer dyes are undergoing safety tests required for FDA clearance. A yellow dye, under investigation, may ultimately replace FD&C Yellow No. 5 and No. 6. A blue dye is still in research development stage.

Food flavorings: The French Academy of Medicine went on record as opposing all practices that lead to the introduction into foodstuffs, for nonphysiological reasons, of foreign substances such as natural aromatic substances or analogous synthetic substances, even when these are reputed to be harmless. The French government permits the use of only seven synthetic food flavorings, which makes France, in this respect, to be the only European country with highly restrictive flavoring legislation. By contrast, some sixteen hundred food flavorings and adjuncts are currently used in the United States. Of these, most have never been tested adequately, and some have never been tested at all. Numerically, food flavorings form the largest single group of food additives.

They pose, as one critic said, "one of the areas of greatest toxicological uncertainty at present."

After the enactment of the Federal Food Additives Amendment of 1958, food additive manufacturers were obliged to determine the safety status of their products. In order to evaluate synthetic flavorings and adjuncts, the food additives committee of the Flavoring Extract Manufacturers Association of the United States (FEMA) collated data collected from their industry-wide questionnaire concerning the identity, composition and uses of these substances. More than a thousand substances were reviewed by an expert panel of toxicologists and pharmacologists who decided whether or not each food flavor additive could be considered as GRAS.

Food flavorings and adjuncts used prior to 1958 could be granted GRAS status in one of two ways. The substance could be evaluated "by experience based on common use in food," or by scientific procedures. Common use, or long-time use of substances in food, has been shown repeatedly to be an unreliable method to determine safety. For nearly seventy-five years, coumarin, a food flavoring, was widely used. Long-overdue tests showed that coumarin caused liver damage in rats and dogs, and it was banned from use. Safrole, a flavoring substance long in use as the chief flavoring ingredient in root beer, was finally tested in long-term feeding with rats and dogs. Safrole was found to be a potent cancer-inciter in the liver, and it was banned from food use.

The review of scientific data by FEMA was also an inadequate method of determining the safety of food flavoring additives in use prior to 1958. In many instances, then as today, information was incomplete or lacking. Testings had been conducted for short-term animal feeding, not for long-term human ingestion. Nor did the data include types of tests that we now recognize as being vital: chronic toxicity, allergenicity, carcinogenicity, mutagenicity, teratogenicity, and other subtle biological damages which can be inflicted on the body by ingested substances.

Even with what appears by our present perspective to be a totally inadequate yardstick for evaluations, FEMA admitted that some one hundred to two hundred food flavoring additives would probably have to be dropped for lack of supporting data or for suspected toxicity. Of some fourteen hundred food flavorings and adjuncts, 191 were covered by FDA "white lists" of official approval; 343 were granted "extension" status; FEMA, on the basis of its own review, decided that 662 of the additives could be considered GRAS. As was mentioned earlier, but bears repetition, additives on the GRAS list automatically enjoy certain privileges. They are exempt from the legal definition of being a food additive. They are not subject to regulations concerning food additives, as long as they remain on the GRAS list. Food flavorings and adjuncts that came into use after 1958 could be placed on the GRAS list only after scientific data were submitted and found acceptable.

Recognizing that toxicity data were woefully lacking in regard to many food flavorings, the FDA conducted studies for acute oral toxicity with rats, guinea pigs and mice for a large number of these substances. Numerous adverse effects were observed, including rough fur and scrawny appearance; diarrhea, soft stools and bloody urine; intestinal irritation, liver damage and hemorrhaging; lacrimation, salivation, gasping and respiratory failure, tremors, depression, ataxia, coma and death.

Flavor additives include compounds with a wide variety of chemical structures and mixtures of variable compositions derived from synthetic sources as well as isolates or extractives from natural products. In later studies, the

FDA tested certain food flavorings and related compounds for subacute and chronic toxicity in rats. Many of the substances caused retarded growth; increased mortality; damage to organs, such as liver, kidneys and heart; stomach ulceration; bone changes; and malignant tumors.

Many essential oils are used in food flavorings. They have been on the GRAS list. In tests for chronic toxicity, some essential oils were shown to be mildly irritating to the mucous membrane of the mouth and of the digestive tract. Their ingestion in large amounts is known to irritate the kidneys, bladder and urethra. In the presence of a preexisting inflammatory condition of the urinary tract, small doses of ingested essential oil appeared to worsen the condition. When essential oils were applied to the mice skins, these substances produced moderate to marked skin hyperplasia. In some cases, areas of necrosis with ulceration, "weeping" of skin, and crusting developed. Some essential oils were found to promote skin tumors. Benign warts and malignant skin tumors appeared in some animals first treated with a known cancer-inciting substance, and then repeatedly treated with an essential oil.

Presently, toxicological data for many food flavoring additives are inadequate or lacking. Since these compounds are considered as valuable trade secrets, they are treated as privileged material. Safety studies, even when done, are not made available from the FDA for public or scientific review.

In 1970, some $150,000,000 worth of flavoring additives were used with foods and beverages in the United States. Flavoring additives were the top sellers among all classes of food additives. The greatly increased demand for flavoring additives has been related to the phenomenal rise in the use of convenience foods. Flavoring additives are used by many food processors in the manufacture of cordials and bitters; breakfast cereals; meats; soup;

margarine and shortening; baked goods; icings and toppings; gelatin desserts, puddings, ice cream and ices; syrups, jellies, candy, chewing gum and maraschino cherries; extracts, pickles, condiments and seasonings.

The presence of added flavoring has not necessarily been stated on the labels of foods and beverages. In view of the FDA's growing appreciation of the fact that many consumers wish to have more specific ingredient labeling, in the future added flavors may be more adequately labeled.

The task is awesome. Dr. Sanford Miller, Bureau of Foods, FDA, in addressing food processors in 1979, reported that the agency is attempting to cope with some system that will give complete ingredient disclosure for flavors. Yet some flavor mixes in foods consist of as many as 600 compounds! Miller reported that the FDA is considering a variety of options. One is to require the industry to print the flavor mixture on the label and identify it with a number, such as "mixture No. 42." Then, in every supermarket, a compendium of these mixtures would be available as reference. A consumer who wanted to check on what was present in a particular mixture would consult the compendium.

While such an arrangement would be costly and cumbersome, it might well encourage food and beverage processors to reduce the number of additives used in the food supply.

Meanwhile, it is best to avoid as much as possible any foods and beverages with added flavorings of unknown ingredients. In view of what is already known, and what still remains to be learned, the safety of some flavoring substances appears doubtful.

Hydrogenated fat: Liquid edible oil, converted into hard fat, is used extensively in food processing. The label may list this ingredient as "hydrogenated fat," "hardened fat," "partially hydrogenated fat" or "partially hardened fat." The degree of hydrogenation varies, but the basic process is the same.

The original edible oil is exposed to a high temperature and placed under pressure. Hydrogen is bubbled through the oil in the presence of nickel, platinum or some other catalyst. The hydrogen atoms combine with the carbon atoms, and the resulting product becomes saturated or hardened. (The new compound bears no relationship to the original oil.) It is then bleached, filtered and deodorized into a pure white, odorless, tasteless, highly artificial fat. It may be processed further to make shortening or margarine.

Hydrogenated fats offer many conveniences to the food industry. They have a high degree of durability, and resist change or deterioration even under poor storage conditions and unfavorable temperatures. "Hydrogenated shortenings can be accompanied by a tenfold increase in resistance to rancidity," asserts the food technologist. These fats do not absorb odors readily. The baking industry finds them "superior to oils in shortening effect and creaming quality." They are well suited for ease and economy of food preparation in the factory, restaurant and home.

The greater the degree of unsaturation in the natural oil, the greater its tendency to respond to the hydrogenation. The heating of the oil ruins its original character, with a destruction of all the vitamins and mineral factors, as well as an alteration of the proteins. The essential fatty acids (EFA) are destroyed or, worse yet, changed into abnormal toxic fatty acids antagonistic to EFA. The synthetic fat forms new molecular structures unacceptable to the human body. This is how Dr. Franklin Bicknell, the author of the prestigious *Vitamins in Medicine*, described it:

The abnormal fatty acids produced by "hardening" are the real worry. The atoms of the molecule of an essential fatty acid are arranged in space in a particular manner . . . but hardening may produce a different spatial arrangement, so that a completely abnormal . . . unsaturated fatty acid is produced. An analogy is ordinary handwriting and mirror handwriting: both are identical but spatially different, so that at best reading the latter is difficult and at worst serious mistakes are made. The same mistakes are made by the body when presented . . . [with the abnormal] EFA. Not only does it fail to benefit by them, but it is deluded by their similarity to normal EFA and so attempts to use them. It starts incorporating them in biochemical reactions and then finds they are the wrong shape: but the reaction has gone too far to jettison them and begin again with normal EFA, so they are not only useless but actually prevent the use of normal EFA. They are in fact *anti-EFA*. They accentuate in man and animals a deficiency of EFA. An analogy is jamming the wrong key in a lock: not only is the lock not turned but the right key also is rendered valueless.

The lack of EFA has been found to be a contributory cause in neurological disease, heart disease, arteriosclerosis, skin disease, and various degenerative conditions such as cataracts, arthritis and cancer.

Dr. Ancel Keys has warned that the unnatural forms of polyunsaturated fatty acids, produced during the hydrogenation processes, have altered biological qualities in the foods. "Where there are double bonds, different configurations of the molecule are found. In natural fats, almost all of the unsaturated fatty acids have the *cis* configuration [atoms or groups of atoms on the same side of the molecule]. The molecule bends backwards on itself at this point. When the lipid chemists hydrogenate these fatty acids, some of the double bond remains but the molecule is straightened out . . . There is much reason to suspect that the *trans* acids [atoms or groups of atoms on opposite sides of the molecule] are biologically less desirable."

As early as 1956, the prestigious British medical journal *The Lancet*, warned in an article on fats and disease that "The hydrogenation plants of our modern food industry may turn out to have contributed to the causation of a major disease." Two decades later, increasing attention is being given to the possible adverse effects of hydrogenated fat on human health, especially as a causative factor in heart disease, atheromas, and cancer. It has been noted that the unnatural fatty acid isomers formed by the hydrogenation process impair the way in which cholesterol is cleared from the system, resulting in high blood cholesterol levels.

In 1979, Mark Keeney, professor of biochemistry, University of Maryland, reported that hydrogenated fats may adversely affect the cell's energy generators as well as red blood cells. Keeney and his associates reported that when *trans* fatty acids are digested, they become part of, and appear to alter, the properties of the body's cell membranes, thus making it easier for cancer-causing chemicals to enter cells. Keeney *et al* suggested that if a causal relationship is established between dietary fat and cancer, then the hydrogenation process as a factor needs careful scrutiny.

Elsewhere, in 1978–1979, Duane G. Wenzel, University of Kansas at the Lawrence campus, reported that *trans* fatty acids produced a variety of effects on cultured heart muscle cells. Among the effects were altered enzyme systems involving fat metabolism, changes in the blood systems, impaired growth, testicular degeneration, and cardiac injury. This continuing study has shown that these fatty acids induce beating heart cells to arrhythmia, which indicates cell injury.

Unfortunately, there has been great confusion of terms, with many individuals thinking that "saturated" is synonymous and interchangeable with "hydrogenated." These terms are not the same. In order to clarify this issue, on March 28, 1978, the FDA issued new regulations for the labeling of fats and oils. The regulations require that the terms "hydrogenated" and "partially hydrogenated" be used instead of the terms "saturated" and "partially saturated." The agency explained that all vegetable oils, whether processed or not, are at least partially saturated, but a partially saturated oil is not necessarily hydrogenated. The terms "hydrogenated" and "partially hydrogenated" describe the chemical process of adding hydrogen to a natural fat or oil in order to make the oil thicker or to turn liquids into solids. The FDA set July 1, 1979, as the deadline for industry to make labeling changes.

The Canadian Health and Welfare, comparable to the FDA, went further in clarifying the issue for consumers. After explaining the chemical modification of fats and oils by means of hydrogenation, the Canadian agency warned consumers that "consequently, it follows that a margarine made from corn oil does not have the same characteristics as the liquid corn oil. For this reason, statements such as 'made from 100 percent corn oil' or 'made from 100 percent vegetable oil' have no importance as far as the polyunsaturated fatty acid content is concerned," since hydrogenation converts a large percentage of oil to mono-unsaturated or saturated.

Americans now eat almost six billion pounds of hydrogenated vegetable oil yearly, in margarines, cooking oils, and shortenings. The sheer volume—some thirty pounds per person annually—deserves far closer examination.

Despite the shocking implications of hydrogenated fats in terms of human health, the process is used almost universally by food processors, and is sanctioned by government agencies responsible for the consumer's welfare. It is difficult, if not virtually impossible, to avoid hydrogenated fats, commonly used in restaurants and bakeries, and in hundreds of consumer food products. They are the fats of margarines and shortenings; in baked and fried goods; in filled and imitation milk; in coffee whiteners; and in peanut butter.

Mannitol: Mannitol is as sweet as dextrose, but being only partly absorbed, yields fewer calories. Mannitol is a sugar alcohol. (*See Sorbitol.*) Mannitol is used in the powdery coatings of some chewing gums, as a release agent. Like sorbitol, mannitol causes diarrhea in relatively small amounts. It accumulates water during its slow passage through the intestinal tract.

Modified food starches: Raw food starches are changed to make them more useful for food processing. They are converted to change their properties regarding texture, resistance to high temperature, cooking behavior, cold water solubility or ability to form gels and pastes.

The food industry uses over 300 million pounds of modified starches annually. These come mainly from corn, sorghum and wheat, and, in more limited amounts, from potatoes, tapioca and arrowroot. Although the basic foods from which the starches are made make good nutritional contributions to the diet, the consumption of large quantities of the starch fraction does not. Modifying the starches renders them even more worthless. These overprocessed foods are high in calories and low in nutritional values. Over-processed starches and sugars supply a major portion of the caloric value of the average American diet, but they are very poor sources of vitamins, minerals, proteins, trace elements and enzymes.

Modified starches, and starches, have many uses throughout the food industry. Some are used as diluents (as in baking powder), as antisticking agents (with bread dough), as absorbing agents (in confectionery sugar), as molding agents (to help cast confectionery pieces into shapes), as bulking agents and fluidizing agents, as coaters, binders, fillers and stabilizers.

In baking, modified starches change the textures of wheat flour used in crackers. The protein content, which needs to be reduced to prevent a doughy texture, can be reduced either by raising the amount of sugar and shortening, or by adding starch. Since starch is less costly than sugar and shortening, starch is used to adjust the flour. Modified starch may comprise from 5 to 15 percent of the flour by weight.

Modified starches are also useful in baking to achieve stability, sheen, clarity and smooth textures for fruit pie fillings, fruit fillings and imitation jellies. They also serve as stabilizers for whipped toppings, icings and cream fillings.

The canning industry uses modified starches for many specialty-type foods, including baby foods, soups, sauces, pie fillings, gravies, vegetables in sauces, chow mein, chili and spaghetti, stews and cream-style corn.

Baby-food processors add modified starch to their products to prevent the development of a watery appearance (unappealing but harmless) that develops in parts of the food mixture that has been in contact with the baby's saliva. This practice is deplorable. The watery appearance may be unaesthetic to the mother, but surely it is unobjectionable to the baby. The practice, however, *is* objectionable. Critics of the baby food companies argue that modified food starch is used for economic purposes, to extend the shelf life of baby food products, thus making them easier to market.

If the use of modified starches was merely an economic issue, it might be viewed as an innocuous practice. However, during the McGovern Committee Hearings on Nutrition and Human Needs, nutritionists testified that the increasing use of modified food starch in baby foods contributes to a decrease in the amount of iron, copper, and other nutrients in the food. Public interest groups pointed out that modified starch may comprise as much as one-fourth of the solids in some baby foods, thus replacing more expensive and more nutritious ingredients, as well as giving the products a deceptive appearance of containing greater contents of solids than they actually do.

Even more important, certain experiments suggested that modified food starches were not well received by infant digestive systems. The critics urged the McGovern Committee that extensive studies be undertaken to determine how effectively infants digest modified starch; some researchers found that infants under six months did not digest at least one type of modified starch efficiently.

Modified food starch is no longer normally digested, according to a study by Thomas A. Anderson, National Research Laboratory, H.J. Heinz Company, conducted in 1969. Anderson reported that modified starches are not digested in an identical manner to native (unmodified) starches, and modified starches raised the blood cholesterol level.

Elsewhere, studies with test animals fed high levels of treated starches displayed some abnormalities, ranging from pulmonary lesions to calcium deposits in the kidneys of some female rats. FDA scientists reported that enlargements of the caecum (a pouch located at the beginning of the large intestine) "is apparently a well-known effect of eating certain modified and unmodified starches. Since the enlargement is not accompanied by pathological abnormalities, it is considered an adaptive rather than a toxic phenomenon."

A number of alkalies and acids are used in converting food starches. These substances, known as starch modifiers or starch-modifying agents, are highly corrosive in large amounts. Their by-products and unreacted chemicals need to be washed out of the modified food starches. SCOGS, in reviewing modified food starches as part of the food additive safety review, in 1978 urged that more studies be developed on the effects of modified food starches on health. The committee evaluated not only the starches themselves, but also the chemical modifications of starches and processes that produce starches with specific functional qualities. One group of starches (distarch glycerols) is produced by cross-linking starches with epichlorohydrin, a substance suspected as a carcinogen. According to SCOGS, the mutagenic properties of epichlorohydrin under specific conditions raise some questions. Use of starches so cross-linked has been discontinued by the food industry until animal feeding studies can be completed. Additional uncertainties were raised, concerning the use of other substances in modifying starches, including the use of hypochlorite oxidized starches, starch succinates, hydroxypropyl distarch, phosphate, hydroxypropyl starch, and distarch propanols.

The safety and suitability of using modified food starches in baby foods had been evaluated and cleared by the National Academy of Sciences in 1970. In view of newer data, modified food starches deserve another review.

Monosodium glutamate is a "flavor enhancer," also known as MSG, and has had a long history of use in food preparation and serving in the Orient. Extracted from seaweed or soybeans, it was used in the Orient to impart a meaty flavor to dishes in which meat, scarce or costly, was used sparingly.

The composition of the MSG now added to more than ten thousand different processed food items in America differs from the Oriental product. It is manufactured from wheat or corn gluten, or from sugar-beet by-products. At present, more than 40 million pounds of MSG are sold annually to add to American foodstuffs, and it has become the most ubiquitous of all food additives. By 1980, sales figures for flavor enhancers are projected to be nearly double those for 1970.

When first introduced in the United States, MSG was used in canned soups. Presently, it is virtually impossible to avoid MSG in any processed food. MSG is found commonly in heat-and-serve convenience foods; meats, stews and meat tenderizers; canned and frozen vegetables; seafoods, fish fillets, clam chowder, codfish cakes and canned tuna; poultry and chicken à la king; almost all canned soups and soup mixes; condiments and seasonings; pickles; salad dressing, French dressing and mayonnaise; imitation maple syrup; potato chips; crackers and other baked goods; candy; baby food; tobacco; and animal feed.

Although MSG was considered GRAS, it should never have been placed on that list, since before the list was drawn up, a medical journal had already reported that MSG had caused eye damage in newborn mice.

Why has the food industry displayed such enthusiasm for MSG? It is a material described as being "much used to step up the indifferent or undistinguished flavor of many canned and processed foods." Frozen-food processors discovered that MSG acted as a color and flavor preserver. A trade release noted that MSG prevented or retarded the development of "warmed over" off-flavors that develop during storage. Canners were told that MSG "helps to maintain the fresh-cooked quality of canned food despite packing operations, long storage, and reheating by the housewife" and that it "restores flavor to your products" and "protects flavor lost through over- and under-cooking." MSG also appears to have an antioxidant quality, since it was advertised to "suppress oxidized flavor which may develop during storage." Food technologists demonstrated its antioxidant effect on hams, bacon, pork sausage, and fatty fish, as well as "acceptability of MSG in frankfurters." MSG offers economic benefits to food processors, since the substance extends the flavor of other foods, and thus allows manufacturers to use smaller amounts of more costly food flavoring ingredients. MSG also has an ability to mask unpleasant flavor notes in certain foods, such as spinach and canned tomato juice. Also, MSG is added to many spice and herb combinations.

Restaurants—from the greasy-spoon diner to the posh establishment—use MSG extensively to "hold the flavor" and overcome "steam-table fatigue" for meats. "Hamburgers or meat patties that cannot be made to order," restaurateurs were advised, "are better if kept on a steam table in a pan that includes MSG in the juices; if stews and creamed chipped beef remain on the steam table, a liberal shake of MSG in small additions of gravy or milk will restore the product's flavor or desirable texture."

MSG has been glamorized for the general public as a flavor enhancer that brings out the natural flavor of foods. It is packaged and sold under several brand names in food stores. Housewives are encouraged to sprinkle it on raw or cooked meats and vegetables before cooking or freezing them. MSG frequently appears either by its generic or trade names under ingredient listings of printed recipes in magazines and newspapers.

Three of MSG's components—namely wheat, corn, and sugar-beet by-products—are common allergens. For those who need to limit or eliminate these substances from their diet, the widespread use of MSG makes this virtually impossible. Even a careful label reading will not always help. Although MSG appears in the ingredient listing on some labels, it does not appear on the labels of others. For example, while MSG, if present, *must* be declared on a soup label, it *need not* be declared on a mayonnaise label.

As early as 1955, *Consumers' Research Bulletin* published a report by a physician who had traced a serious allergy affecting a woman and her son to MSG. Within a half hour after eating meals of excellent food prepared at home, as well as meals eaten at fine restaurants, they developed acute distress resembling gallbladder trouble. Their symptoms included epigastric fullness, belching, distention and marked upper abdominal discomfort. The reactions were traced to the use of MSG.

In 1968, Dr. Robert Ho Man Kwok, writing in *The New England Journal of Medicine*, reported a strange syndrome that he personally experienced in restaurants serving Northern Chinese food. About fifteen to twenty minutes after eating, he developed "numbness at the back of the neck, gradually radiating to both arms and the back, general weakness and palpitations." Dr. Kwok added that some of his Chinese friends, both medical and nonmedical, also complained of similar symptoms after eating Chinese food.

Dr. Kwok's description elicited other reports of what came to be known as Chinese Restaurant Syndrome (CRS). Many physicians confirmed similar experiences, and some included symptoms of "profuse, cold sweat," "tightness on both sides of the head," "felt as though, at any second, the sides of my head would burst," "pounding, throbbing sensation in the head, viselike."

Researchers found that MSG was related to the CRS symptoms, and it then became publicized that the syndrome was already well known to allergists and Chinese restaurateurs. Several teams of researchers, working independently of one another, established a relationship between the CRS symptoms and the ingestion of MSG in food.

To find out whether MSG caused more than mere transitory symptoms, experiments were conducted with animals. MSG, injected under the skin of newborn mice, induced retinal lesions and hypothalamus damage. MSG, fed to newborn mice, induced both hypothalamus damage and brain damage.

MSG-treated adult animals showed stunted skeletal development, marked obesity and female sterility. On the basis of these results, pregnant women were advised to avoid the use of MSG until further data were available.

Further adverse experimental data were reported, before a review was completed by the AMA's Council on Foods and Nutrition. Brain lesions had occurred in every species of experimental animals tested with MSG, including mice, rats and rabbits. The next species tested was a primate: the rhesus monkey. Brain lesions occurred in infant rhesus monkeys treated with a relatively high dose of MSG, injected under the skin. Nerve-cell deaths occurred in seven out of nine baby animals with about five times the amount of MSG used in a 4½ ounce jar of some commercial baby foods. The researchers commented: "The lack of symptoms in the primate during the time when a small percentage of its brain cells were being destroyed is evidence of a subtle process of brain damage in the developmental period, which could easily go unrecognized were it to occur in the human infant under routine circumstances." MSG appears to damage the brain of newborn animals of several species due to the absence of an effective blood-brain barrier in the very young.

As a result of data from these various experiments, leading baby food manufacturers voluntarily withdrew MSG from their products pending a clearer understanding of the mechanism of injury to the developing brain.

Why was MSG ever put into commercial baby foods? It was added to vegetable and meat mixtures to make them more palatable to mothers who sampled them. Nutritionist Dr. Jean Mayer claimed that MSG was added to commercial baby foods "to disguise the fact that there was less meat and more starch" in these products than formerly. He added, "The MSG was a risk added to a disadvantage."

Although MSG may no longer be found on the labels of baby foods, a substance closely related to it has been substituted. Hydrolized vegetable protein, also called protein hydrolysate, is now added. When protein is hydrolyzed, the individual amino acids are released in free form. One of them is MSG. Protein hydrolysate, like MSG, has been shown to damage the infant mouse brain.

In 1976, as part of the GRAS list review, SCOGS prepared for the FDA a tentative evaluation of the safety of monosodium glutamate and other glutamates as food ingredients. The group reviewed reports and studies on the absorption, metabolism and excretion of MSG; human consumption levels; reports of neurological and endocrinological disturbances, decreased body weight, organ atrophy, and other observations in test animals; allergic reactions, behavioral overactivity and disorganization in humans. The group concluded that answers to questions of possible risk to the public required more definitive studies of the pharmacodynamics and toxic thresholds for MSG and other glutamates, as well as the influence of route of administration, age, including prenatal, species, and strain of laboratory animal, and relationship to other food components. "Whether glutamates ingested in amounts similar to those added to foods can produce significant, untoward effects in man has not been established; thus the acute and long-term biological and behavioral effects of both food additive amounts and larger amounts should be studied. Questions concerning protective mechanisms against glutamate toxicity such as intestinal, hepatic, retinal, blood-brain, and placental barriers need answers. Studies should consider the development of, and changes in, these mechanisms in different species, including man, thresholds, and effects of abnormal states." Thus, the SCOGS review, as so often happens, failed to make any determination; in view of lack of information, and equivocal results, the group concluded that with the available information, "there is no evidence" that glutamates demonstrate or suggest "reasonable grounds to suspect a hazard to adults when used at the levels and in the manner now practiced." SCOGS also deemed that "the evidence is insufficient to determine that the adverse effects reported are not deleterious to infants" when glutamates are added to infant formulas and/or commercially prepared strained and junior foods. The group recommended, as is commonly done, that because of uncertainties, additional studies should be conducted.

The SCOGS review did indicate some concern about possible harm to infants. "Despite the fact that the brain damage reported in neonatal [newborn] mice, rats and monkeys by some research groups could not be confirmed by other groups," SCOGS concluded that these biological assaults are real and reproducible, even if their significance is not yet determined. "Based on evidence from laboratory animal studies, it has been postulated that the hypothalamic neuronal damage which could be suffered by an infant exposed to a toxic dose of MSG would be a 'silent lesion' whose manifestations would probably not appear until later life in the form of behavioral or endocrine dysfunctions."

As a result of the SCOGS review, the FDA took extraordinary action, by giving MSG "interim" GRAS status. In view of a plethora of studies incriminating MSG, the FDA *should* have removed MSG from the GRAS list, since it was certainly no longer "generally recognized as safe," even though many questions were unresolved. Among them, the FDA hoped to gather more data on allergic reactions to MSG, since it was reported that 25 percent of the population suffers adverse reactions to this substance.

In 1977, two days of hearings were held on the safety of MSG and other glutamates, with much conflicting testimony from experts, industry, and other interested parties. Dr. John Olney, the scientist who first reported lesions on the brain structures of unborn primates due to MSG, introduced new evidence at the hearing. Olney claimed that glutamates, in smaller doses than those considered to be toxic, could induce changes in the amounts of growth hormones produced, especially increasing secretion of testosterone in the newborn. This evidence suggested that amounts of MSG, considered lower than toxic levels, could nevertheless have a cumulative effect on sexual development. Olney speculated that low doses could cause hormonal disturbances in young animals.

Testimony was also given by Dr. Liane Rief-Lehrer, associate professor, Department of Ophthalmology, Harvard Medical School, who tried to quantify the prevalence of allergic reactions to MSG. In one informal survey she undertook with 300 school-age children, about 20 percent of them had experienced reactions similar to those of CRS, already described. In another informal survey of 1,500 adults, mostly college-educated, about 30 percent experienced CRS. Dr. Reif-Lehrer referred to an earlier study, in 1972, which stated "culinarily acceptable levels of MSG caused reactions in a relatively small number of people, but when the level of glutamate ingested was of the order of 5 grams [the average amount in wonton soup in a Chinese restaurant], 30 percent of the individuals tested responded adversely."

In 1978, SCOGS presented its final report on glutamates to the FDA, which in essence, repeated the tentative report: no health hazards to adults at present levels, but additional studies were recommended. The FDA again took extraordinary action. The agency refused to accept the final report on the GRAS status of MSG. This refusal represented the first time that the agency failed to accept a GRAS review evaluation from FASEB.

The FDA raised questions about new evidence which had not been considered in the SCOGS review. The agency questioned conflicts raised by the final report. Serious questions have been raised about the legal authority of the FDA and SCOGS concerning food additive evaluation, and subsequent regulatory actions. Equally important, serious questions have been raised about the scientific knowledge and implementation of recommendations.

Specifically, the points of scientific controversy about MSG included acute toxicity at predictably possible exposure levels, and conflicts in the available scientific data. There was also dispute over the methodology employed in some studies, and problems with interpretation of similar data by scientists of opposing views.

The six areas of controversy over MSG's safety centered in the following:

—The production of hypothalamic brain lesions in laboratory rodents after MSG had been given.

—The production of other physiological changes in test mice, gerbils, guinea pigs, and other animals including humans.

—The production of hypothalamic lesions in primates following administration of MSG.

—The existence of a "blood-brain barrier" mechanism that acts to protect the body from excess glutamate transfer to the brain, where it could be damaging.
—The existence and prevalence of allergic reactions to MSG by some individuals, commonly referred to as the "Chinese Restaurant Syndrome."
—The ultimate question of the safety of MSG as a food additive at present and future use levels.

The FDA requested that SCOGS examine new information on MSG safety that was not available when the panel made its report. Some of the new information was presented at a symposium in May 1978, sponsored by the International Glutamate Technical Committee, a food industry scientific group. As could be expected, the general conclusion reached by the majority of the participants was that MSG presents no health hazard to adults, even at high consumption levels.

But in October 1978, Dr. Arthur D. Colman, University of California Medical Center, San Francisco, reported psychiatric reactions in an adult to MSG, including "profound depression, drawn facies, motor slowing, doubt-ridden, gloomy fantasies, and occasionally unprecipitated outbursts of rage" beginning 48 hours after exposure to MSG and lasting for as long as two weeks. Although there were no frank hallucinations, the individual felt that people were looking "strange and ominous" with corresponding feelings of paranoia in otherwise routine situations. Colman also described MSG reactions in a nine-year-old boy who displayed typical acute reactions to MSG, along with bladder and bowel incontinence and intermittent hyperactivity, all of which disappeared on an MSG-free diet.

The FDA requested SCOGS to review Colman's claims as well as the newer information presented at the MSG symposium. After the FDA gathered all the accumulated data, it would make them available, and solicit public comment. Pending review of all data, the FDA would defer further action on its review of the GRAS status of glutamates, including MSG.

Saccharin: The nonnutritive sweetener, saccharin, and its ammonium, calcium and sodium salts, were synthesized in 1879, and have been long used in foods as well as in drugs. Saccharin had qualities that appealed to the food industry—being intensely sweet, cheap to manufacture, and easy to handle. Saccharin-sweetened products flooded the market. Saccharin was used as a sugar substitute in candies, soft drinks and bakery products. Canners used it, and one processor reported that his company saved $4,000 in one year by sweetening canned corn with saccharin instead of sugar. Dr. Harvey W. Wiley, chief of the agency that later became the FDA, retorted: "Everyone who ate that sweet corn was deceived . . . he was eating a coal tar product totally devoid of food value and extremely injurious to health." Dr. Wiley had branded saccharin, as early as 1913, as being "a noxious drug, and even in comparatively small doses, it is harmful to the human system." Wiley was unsuccessful in keeping saccharin out of the American food supply by a strange quirk. He presented data to President Theodore Roosevelt demonstrating the harmfulness of the substance. But Roosevelt, who had been put on saccharin by his own physician, refused to accept Wiley's evidence.

Elsewhere, food use for saccharin was more strictly regulated. As early as 1890, the Commission of the Health Association in France decreed saccharin harmful and forbade its manufacture or import. Eight years later, the German government limited its use, and expressly banned it from all food and drink. Similar action was taken in Spain, Portugal, Hungary and other countries.

From time to time, articles and letters in medical journals reported diverse untoward effects of saccharin, including interferences with normal blood coagulation, blood sugar level and digestive functions, and the production of urticaria.

As early as 1951, three scientists within the FDA reported that saccharin at certain levels showed a high incidence of unusual combinations of cancers. But the FDA chose to ignore the report, and it was not known publicly until the cyclamate affair in 1969.

In the wake of the cyclamate ban, new attention was given to saccharin. Dr. George T. Bryan, a tumor expert and cancer researcher, reported to the FDA that by implanting a saccharin pellet in the urinary bladder of mice, bladder cancer was induced in 47 percent in one group and in 52 percent of another group. This same technique of bladder pellet implant had been used in some of the tests that had demonstrated the carcinogenic property of cyclamates. Dr. Bryan admitted that although direct cancer hazard to man from saccharin was not yet established at that point, he was "very suspicious." He added, "It may take many years before it is known exactly how dangerous the subject is, and until then its use should be restricted to those who need it for medical reasons."

Assessing safety test data for saccharin has been especially complicated. Some past studies were poorly designed. The tests had been conducted with impure test substances, at times on unhealthy animals, and nearly all of the early results are of unknown relevance to humans. A scientist, attempting to evaluate saccharin safety tests based on past experiments, described the effort as "an education in obtuseness—poor data derived from poor experiments." He added, "This is a job for a philosopher, not a scientist."

One of the many unresolved problems with saccharin arose when more recent tests induced tumors in some test animals. Were the tumors cancerous or benign? Examination of this question began in October 1969, when the FDA requested a review by the NAS–NRC, which was expected to be completed in approximately two months. In January 1972, the FDA restricted saccharin's use while the safety review was being completed. In September 1972, an FDA spokesman announced that a few more months would be needed. In May 1973, the FDA announced that evaluation was being conducted. Then, it was announced that the review would be completed by the end of June 1973. By September 1973, the panel of scientists was still evaluating the evidence and awaiting new test results. The NAS–NRC announced that its report would probably be finished in the autumn of 1973. Actually, the report was issued the following year, and it concluded that evidence had "not established conclusively whether saccharin is, or is not carcinogenic when administered orally to test animals."

Subsequently, substantial new evidence was generated, and a cascade of data and official actions ensued. On March 9, 1977, the Canadian government banned saccharin, as a result of earlier saccharin studies. The Canadian tests had shown that a five percent dietary level of saccharin increased the incidence of bladder tumors in male rats, especially in the second generation. The Canadian investigators concluded that saccharin is a potential cancer-causing agent. The Canadian National Health Protection Branch, after consultation with the Canadian Medical Association, the Canadian Diabetic Association, the Canadian Dental Association, and the Registrars of Pharmacy, decided to remove saccharin as an additive in processed food.

Prompted by the Canadian study, in March 1977, the FDA announced its intention to ban saccharin under the general safety requirements of the Food Additives Amendment of 1958, and the Delaney anticancer clause to this amendment. On April 15, 1977, the FDA announced its intention to revoke the interim food additive regulation which permitted saccharin as an ingredient in prepackaged foods; review applications for marketing saccharin as a single-ingredient drug available without prescription; ban saccharin as an inactive ingredient in drugs; prohibit saccharin in cosmetics likely to be ingested; and prohibit saccharin in veterinary drugs and animal feed.

These proposals met with sharp criticism, largely from affected segments of industry and users of saccharin-containing items. Millions of dollars were spent by the Calorie Control Council, a saccharin-vested interest group, to convince the general public that saccharin should not be banned from use.

This backlash led the Subcommittee on Health and Scientific Research of the Senate Committee on Human Resources to throw the controversial issue into the lap of the Office of Technology Assessment (OTA) in March 1977, to study the technical basis of the FDA's proposal. OTA was to report back within 60 days.

On June 7, 1977, the essential findings of the OTA were presented and discussed at the hearing held before the Subcommittee on Health and Scientific Research of the U.S. Congress. The OTA report, issued in October 1977, concluded that "laboratory evidence demonstrates that saccharin is a carcinogen," and that "evidence leads to the conclusion that saccharin is a potential cause of cancer in humans." Furthermore, OTA noted that although a nonnutritive sweetener such as saccharin is "of perceived psychological benefit" to many people, whether or not the use of a nonnutritive sweetener leads to measurable health benefits "has never been tested."

Both public and congressional concern about the fate of saccharin mounted. Numerous bills were introduced in both houses of Congress to stay FDA's plans for banning saccharin. On November 3, 1977, The Saccharin Study and Labeling Act was enacted, requiring the

Secretary of Health, Education and Welfare to ask the NAS once again to conduct "studies concerning toxic and carcinogenic substances in foods, to conduct studies concerning saccharin, its impurities and toxicity, and the health benefits, if any, resulting from the use of non-nutritive sweeteners including saccharin." The act prohibited implementation of the proposed ban on saccharin for a period of eighteen months, but required that food products containing saccharin bear a warning on the label. Also, NAS was requested to study and develop policy recommendations concerning the safety of the nation's food supply—a demand that would broaden the issue considerably, and cause further delays in decisive actions.

While this eighteen-month moratorium on the ban was in effect, other events were taking place with saccharin. On January 25, 1978, the FDA and the National Cancer Institute (NCI) announced plans to conduct a nationwide study on the possible role of saccharin in causing bladder cancer in humans. Also, the FDA negotiated with the NAS for a number of other saccharin studies, required by the moratorium legislation, to assess the health benefits and risks.

The following month, the FDA announced plans to require a warning label on saccharin-containing foods. Three months later, the agency announced similar plans, to require warning posters in stores. Such warning statements for saccharin are about as ineffective as they are for tobacco.

Finally, on November 4, 1978, the panel of NAS released its fifth review of saccharin since 1955. The report held few surprises. The panel did conclude that saccharin posed a potential carcinogenic risk to humans, albeit a comparatively low risk. This was the first time the NAS panel was willing to acknowledge this fact in its report. The panel also concluded, for the first time, that saccharin promotes the development of cancer initiated by other substances.

Despite NAS refusal to make a precise risk assessment, the panel did target several groups whose risk of cancer from saccharin consumption is greater than others. *One-third of children under ten years of age consume saccharin-containing food products. Based on consumption per body weight, children under ten have the greatest intake of saccharin of any age group.* Since 1972, saccharin consumption among children has jumped 160 percent, according to market data, and it is mainly from soft drinks. Children are particularly vulnerable. The amount they consume, relative to body weight, makes them a group with the greatest exposure.

While most of the thirty-seven-member NAS panel succeeded in clearing the way for federal approval of saccharin, about a dozen issued a "minority statement" dissenting in the strongest possible terms from the report's endorsements, and even made some recommendations that were totally at odds with the majority. Dr. Frederick Robbins, Dean of the School of Medicine, Case Western Reserve University, who headed the NAS panel, recommended that saccharin should be phased out, in an orderly fashion, within three years.

After all the saccharin reviews, the status of this substance remained unclear. The Senate Health Committee had thrown the problem into the lap of NAS, and NAS returned it to the lap of Congress. Although the NAS panel had little disagreement about the scientific aspects of saccharin, ultimately the group decided it was up to the politicians rather than the scientists to decide on the fate of saccharin. The frenetic activities revolving around saccharin finally led to a far larger debate over food safety policy, the concept of benefit versus risk, and the Delaney Clause. A total revision of the food safety laws was proposed by NAS.

On May 23, 1979, the moratorium on saccharin banning expired. The FDA made no attempt to revive its proposal for a ban, and Congress extended the moratorium for another two years. From this action, it is clear that the saccharin issue is not a scientific one, but rather political in nature.

Silicates: Silicates are used as anticaking agents in dry powdery foods such as table salt, vanilla powder, vanilla-vanillin powder, garlic powder and dry soup mixes; and in dietary supplements to render them free flowing. Some are used in baking powder and with dried egg yolk. They also serve as release agents and dusting agents with confections and chewing gums, and as a flavor fixative on polished white rice.

Several forms of silicates are used for these purposes: hydrous aluminum calcium silicate (calcium aluminosilicate) which occurs in nature in at least twenty-two minerals; calcium silicate, which occurs in at least twenty-two minerals; hydrous magnesium silicate (silicate of magnesium), well known as talc, of which several varieties are known and which occurs in at least four minerals; sodium aluminosilicate (sodium silico aluminate), which occurs in at least three forms as aluminum silicate, and is found in at least ten other aluminum silicate minerals; sodium calcium aluminosilicate (hydrated sodium calcium silicoaluminate); and tricalcium silicate. It is apparent that silicates are found in a variety of forms, from many sources.

One of the silicate food additives deserves special attention: talc, or hydrous magnesium silicate. Some forms of asbestos are *also* hydrous magnesium silicate. Talc mineral beds may contain several forms of asbestos, and some talc powders may be contaminated with asbestos. The differences between the two minerals are not in their chemical compositions but rather in their structures. Asbestos is fibrous, while talc is flaky or granular.

Frequently, commercial talc is contaminated with fibrous silicates which, in part, can be classified as asbestos. Thus, as can be readily understood, the asbestos-contamination of talc presents a hazard whenever talc is used as a food additive.

The hazard of *inhaling* asbestos fibers is well established. Once these incredibly small particles are taken into the lungs, they remain there and provide a lifelong potential for the development of cancer.

The hazard of *ingesting* asbestos fibers, or asbestos-contaminated food and beverage, is a subject largely uninvestigated. The few fragments of information that have been reported indicate this should be an area of vital concern.

California talc is frequently used to coat polished white rice, which is shipped to Japan and consumed in large quantities in some areas. When analyzed, California talc has been found to contain from 20 to 40 percent tremolite (calcium magnesium), a form of asbestos.

Dr. Reuben R. Merliss suggested that the ingestion of this asbestos-contaminated talc on the rice might be a factor in the high incidence of stomach cancer in Japan. (Many other factors have been suggested.)

Although thorough rinsing of the rice may remove a great deal of the asbestos-contaminated talc, Dr. Merliss found that the process failed to free the rice completely from the coating. He found that rice that had been cooked, canned and sold in Japan still showed as much as 1.1×10^4 asbestos-form fibers per gram of desiccated cooked rice. Dr. Merliss noted that talc is usually considered so innocuous that specific inquiry or examination must be made to determine its presence as a food additive.

Asbestos fibers may become accidental additives in foods and beverages in ways other than as a talc-contaminant. At times, asbestos fibers have been found to contaminate fruit juices, acids and beers from filtering processes.

Two scientists in the Canadian Food and Drug Directorate analyzed twelve brands of Canadian beer, wine, sherry and port, six of American beer, six of European wine, vermouth and sherry, and two of South African sherry. All samples were found to contain asbestos fibers. The fibers, so incredibly tiny, were visible only through an electron microscope. The researchers suspected that such materials could pass through filtering systems of drinking water supplies as well. To confirm this, the two scientists analyzed samples of tap water from three major Canadian cities with modern filtration systems: Ottawa, Montreal and Toronto. They found from 2.0 to 4.4 million asbestos fibers per liter.

The source of this contamination is unknown. Major asbestos mines are located in Quebec, but a more likely source appears to be industrial pollution. The scientists found high levels of asbestos fiber in melted snow in the Ottawa area.

The Canadian scientists believe that the levels of asbestos fibers they found in the alcoholic beverage samples were, if anything, conservative. "Many asbestos fibers were hidden in clumps of debris such as mineral matter in water and diatomaceous earth which is used as a filtering aid in beer. As a result, the number of fibers reported . . . must be considered a minimum estimate of those actually present."

While orally ingested asbestos is believed to cause gastrointestinal cancer, the Canadian researchers suggested the need for more knowledge to determine "the extent to which the fibers penetrate the walls of the digestive tract, the degree to which they may be transported to other organs of the body and the biological effects of their residence in various tissues." The subject of asbestos-contamination of the environment, in air, foods and beverages, serves as a good illustration of the complexities involved with the subject of food additives.

Sodium carboxymethylcellulose, also known as cellulose gum or carboxymethycellulose, are compounds described as "undigestible materials" that are "often substituted for normal food ingredients in an effort to improve texture and keeping quality or to lower calorie value and cost." The United States Public Health Service (USPHS) warned that "if carried to excess, such modification may have serious nutritional consequences to certain segments of the population," and noted that such nonnutritive additives have "physiological potentials which are . . . incompletely explored at the present time."

Sodium carboxymethylcellulose is used as a stabilizer, thickening agent, and suspending agent with many foods. As a stabilizer in ice cream and sherbet, this additive prevents large ice crystals from forming. In chocolate-flavored milk, it prevents the particles of chocolate from settling to the bottom of the drink. In low-calorie canned fruits sweetened with nonnutritive sweeteners, it gives mouth feel by acting as a bodying agent. It is also used in beverages; confectionery; baked goods; icings and toppings; pressure-dispensed whipped cream; syrups for frozen products; frozen variegated mixtures; cream cheese, cottage cheese and cheese spreads; French dressings and salad dressings; artificially sweetened jellies and preserves; and in relishes, mustard and pickles.

Sodium carboxymethylcellulose is used by many bakers and dairymen even without their being aware of it. A dairy may buy an ice cream stabilizer from a stabilizer manufacturer, for example, and it may contain sodium carboxymethylcellulose along with gelatin, vegetable gums and seaweed derivatives. Similarly, a bakery may buy fillings and meringue mixtures that include sodium carboxymethylcellulose. Neither the dairy products nor the baked goods will necessarily have this information on the label listing of ingredients.

Sodium carboxymethylcellulose is considered GRAS. Feeding rats this substance produced negative results. But when, during a two-year study, this compound was injected under their skins, it produced sarcomas. It induced arterial lesions similar to those produced by high cholesterol in blood. One intravenous injection resulted in the loss of appetite and weight, progressive loss of body fat, and enlargement of the liver and spleen. Visible to the naked eye, there were thickenings in the arteries, with numerous and extensive splits and tears along the artery walls. Less apparent, but present, were tears penetrating into the deeper layers of the arteries, blood vessels and heart muscles. Early signs of thrombosis were detected. The mechanism seemed to interfere with nutrition of the blood-vessel walls, and caused secondary destructive changes, especially in the muscles and blood.

Although originally, the FDA had rejected a proposal for the inclusion of sodium carboxymethylcellulose in foods, the policy was later reversed. Dr. Wilhelm C. Hueper, former chief of environmental cancer research at the National Cancer Institute, considered this reversal as "ill-considered recommendations that the FDA continues the use of . . . carboxymethylcellulose as a food additive." Dr. Hueper had worked with these substances experimentally, and had found that carboxymethylcellulose produced a significant number of tumors and "cancerous responses" in rats. He also pointed to the work of Dr. G. Jasmin of Montreal who, as a research associate of the Canadian Medical Research Council, had reported carboxymethylcellulose as a cancer hazard as early as 1961. In 1969, the researchers at the National Cancer Institute furnished fresh evidence concerning the hazardous properties of this substance. They discovered that it was the probable cause of a cystic tumor in a rat's kidney.

Dr. Hueper stated that sodium carboxymethylcellulose should be forbidden as a food additive. The substance does not *have* to be used. In an interview, Dr. Hueper said, "I am sure that the food industry can get along without it. If carboxymethylcellulose were prohibited, it would not endanger the industry's existence. There are a lot of other natural mucilages that can be used as emulsifiers and suspending agents."

Sodium nitrite and sodium nitrate: These compounds are extensively used as preservatives and "color fixatives" in cured meats and meat products as well as in certain cured fish. There have been instances of sodium nitrate being added illegally as an adulterant to fresh meats, and also incidents of illness and even death after persons have eaten fresh fish to which sodium nitrite had been added illegally and in recklessly large amounts.

Human toxicity for these substances has been well established. Dr. Harvey W. Wiley refused to approve their use in cured meats, although they were never expressly forbidden by official ruling. Wiley felt that no matter how small the quantities were, their use in food products was undesirable.

Data from human experience and animal experiments have shown adverse effects from sodium nitrite and sodium nitrate. Of the two compounds, sodium nitrite is the more toxic. However, sodium nitrate readily converts to sodium nitrite.

Recurrent and moderate to severe arthritic symptoms were self-diagnosed by a physician after he ingested foods containing sodium nitrate.

Sodium nitrite, administered for several days to chickens, inhibited their normal ability to store vitamin A, or carotene, in the liver, even when the diet was well supplied with these nutrients. Sodium nitrate showed similar inhibiting properties.

Sodium nitrate and sodium nitrite, fed to cattle, sheep

and rats, induced similar decreased liver storage of vitamin A and carotene. Some animals showed vitamin A deficiency symptoms.

An Israel research team reported that sodium nitrite, fed to rats regularly in their diet, induced permanent epileptic-like changes in their brain activity. Doses were only slightly higher than those that would be ingested by persons consuming large quantities of frankfurters, cured meats and sausages.

The problem of toxicity for sodium nitrite has become far more complex since it was discovered that nitrite can combine with other substances and form nitrosamines, a class of compounds termed "the most potent cancer-causing agent known to science." The sequence of development chronologically is as follows:

—In 1956, a nitrosamine compound was found to cause liver cancer in rats.

—In 1958, the Delaney Clause banned from use as food additives any substances known to be carcinogenic in humans or in animals, using appropriate tests.

—In 1959, Canada prohibited the addition of potassium and sodium nitrate to smoked fish because of toxicity.

—In 1970, Drs. Samuel S. Epstein and William Lijinsky reported a link between nitrite and nitrosamine formation, and recommended that the nitrite level should be drastically reduced or eliminated completely as a food preservative and color fixative. Their recommendation was based on significant studies indicating that nitrites may be carcinogenic, mutagenic, and teratogenic. The question posed was this: Is it possible that humans may synthesize nitrosamines within the body? It was known that nitrates and nitrites, found in common foods, in combination with secondary amines also present in some common foods, may react in the acidity of the digestive tract to form nitrosamines. The high concentration of nitrates and nitrites permitted in many foods as additives leads to an almost constant presence of nitrite ions in the stomach.

—In 1971, the nitrite-nitrosamine link received a good deal of attention during Congressional Hearings on Regulations of Food Additives and Medicated Animal Feeds. Although experts testified that nitrites at the high levels used, mainly for "cosmetic purposes," were hazardous, government officials saw "no imminent hazard." However, work was begun at Massachusetts Institute of Technology (MIT), and it was discovered that nitrites caused lymphatic cancer in laboratory rats.

—In 1972, the FDA and the USDA promised to establish a "panel of distinguished scientists" to evaluate the significance of the potential carcinogenesis of nitrates and nitrites. This panel, once formed, was to meet over the next few years, with a good deal of foot-dragging, lack of decisions, and repeated requests for more time and studies. At the same time, public interest groups demanded that prompt actions be taken to eliminate nitrates and nitrites from the food supply, and that safe alternatives be developed.

—In 1973, the use of ascorbates was being investigated to block nitrosamine formation in nitrite-treated cured meat products.

—In 1974, two Japanese researchers reported that two nitrosamine compounds, given in drinking water to rats, produced, in one case, hepatomas, and in the other case, papillomas in the esophagus.

—In 1977, the *Journal of the American Medical Association* reported that "since no species is known to be resistant to carcinogenesis by the N-nitroso compounds [nitrosamines], there is good reason to believe that humans are also susceptible." The periodical made reference to the fact that ascorbic acid, added to the diet, reduced by half both the carcinogenicity and the N-nitroso compound level in the feces.

—In 1978, Dr. Paul Newberne's studies at MIT, begun in 1971, reported that malignant lymphoma was induced by nitrites in laboratory rats. Although there had been some discussion about a possible nitrite ban, no definite action had been taken by federal regulatory agencies. After Newberne's disclosures, there was talk merely of a "phaseout" of nitrite.

—In 1979, a new study conducted by Dr. Newberne for the FDA was even more forceful, since it directly linked nitrite with cancer. According to the study, 13 percent of laboratory animals fed nitrites developed cancer, while only 8 percent of the animals in a control group not fed nitrites developed cancer. The 5 percent difference is statistically significant. Formerly, the main concern was about nitrosamine formation; now that the nitrites themselves were directly linked to cancer, the government had several courses of action possible. The FDA could ban nitrites by deciding that nitrite-containing products were deemed to be adulterated. Or, it could order further reductions in nitrite levels. Or, it could require warning labels for nitrite-containing products. Unfortunately, the Delaney Clause was not applicable, since nitrites had been long in use before the 1958 clause was enacted into law.

In response to a request for an immediate ban, federal officials refused, claiming that it was essential "to assure the safety of the nation's food supply." Thus, twenty-three years after a first demonstration of the carcinogenesis of nitrosamines, the regulatory agencies have failed to act. Instead, they requested an opinion on the possible illegality of a "phaseout" of nitrites from the food supply, since a three-year phaseout has been suggested. And even 1982 may not be the final date, if additional delays are thought to be needed.

Arthur M. Sackler, M.D., publisher of *Medical Tribune*, commented on "Violating the Law with Impunity" by the FDA and the USDA in respect to the nitrite issue. These regulatory agencies have "failed to act to get a carcinogenic additive completely out of our food supply. . . failed to seek significant reduction in carcinogenicity levels of one of the most potent potential human carcinogens, which, according to uncontested scientific evidence, could be cut in half by the simple addition of ascorbic acid. Why has there for years been a delay in protecting Americans from a carcinogenic food additive? . . ."

Sorbitol: Sorbitol is one of many sugar alcohols, which contains calories, and is metabolized by the body to produce energy. Although sugar alcohols may be more poorly absorbed than normal sugars (from cane, beet, corn, etc.) and absorbed more slowly, nevertheless they are metabolized. Other sugar alcohols include mannitol and xylitol. Promotion of these sugar alcohols, as "low-calorie" alternatives to common sugars is somewhat deceptive. They are neither low-calorie, nor can they be viewed as desirable sugars for diabetics since they can and do lead to the production of some blood sugar. An FDA regulation, scheduled to become effective in 1980, requires manufacturers to state on labels that inclusion of these sugar alcohols (also called "rare" sugars) does *not* mean that the food or beverage is "low-calorie" or "reduced calorie."

Sorbitol is used in chewing gums, mints, candies, and dietetic ice cream. Sorbitol is also used by food processors to promote retention of the original food quality during storage and shipment, or to improve the texture of food since sorbitol can function as a crystallization modifier, humectant, softening or plasticizing agent, sweetness or viscosity controller, or rehydration aid. For these purposes, food and beverage processors use sorbitol in baked goods, frozen dairy products and milk products, poultry and fish products, nonalcoholic beverages, meat products, frostings, snack foods, processed fruits, nut products, fats and oils, gelatin puddings, alcoholic beverages, sweet sauces and seasonings and flavors. In fact, the use of sorbitol has escalated so sharply that the SCOGS' panel was unable to arrive at any estimate of the average daily intake. There was such discrepancy of figures, the data were deemed to be totally unreliable.

Studies have shown that young children experience diarrhea when they eat sorbitol-contained candies. A laxative threshold has been established in adults. Rats fed 16 percent sorbitol in their diet for nineteen months showed a tendency to become hypercalcemic after one year. Some of the animals had bladder concretions and a generalized thickening of the skeleton.

In reviewing sorbitol, SCOGS noted that this sugar alcohol begins to have a laxative effect at levels that are only about twice the average adult intake level, and about equal to the maximum adult intake level. *The average consumption levels of children in the age groups six to eleven months and twelve to twenty-three months are now estimated to be close to, or even in excess of, those capable of exerting a laxative effect.*

SCOGS was concerned that the actual consumption of sorbitol may be considerably higher than average consumption, in certain segments of the population. The panel recommended that "in order to conduct a more accurate survey on the intake of substances used in food processing" for substances such as sorbitol "food consumption data collected specifically for this purpose are needed." Despite the lack of data, SCOGS concluded that no evidence demonstrated that sorbitol as a food ingredient constitutes a hazard to the general public when used at current levels, or at levels that might be reasonably be expected in the future. This conclusion flies in the face of facts. What is not looked for, will not be found.

Sulfur dioxide and sulfites: Sulfurous compounds have long been used as preservatives, antioxidants and antibrowning agents with many foods and beverages, except with meats or in foods recognized as sources of vitamin B_1 (thiamin). Sulfur dioxide has been used as a bacterial inhibitor in wine; in corn syrup and table syrup; in imitation jelly; with dried fruits, brined fruits, glacéed fruits and maraschino cherries; in beverages; in dehydrated potatoes; in soups and condiments. Sodium sulfite has been used as a bacterial inhibitor in wine, brewing and in distilled-beverage industries; as an antifermentative in sugar and syrup industries; as a preservative for fruit juices, meat syrups and vegetable juices; as an antibrowning agent in cut fruits, frozen apples, dried fruits, prepared fruit-pie mixes, and peeled potatoes. Potassium bisulfite, also called acid potassium sulfite or potassium hydrogen sulfite, has uses similar to sodium sulfite, and is also used in ale and beer. Sodium bisulfite, also known as acid sodium sulfite, or sodium hydrogen sulfite, or sodium acid sulfite, has uses similar to sodium sulfite in ale and beer; and it has uses similar to sulfur dioxide in fruit-pie mixes. Potassium metabisulfite and sodium metabisulfite have uses similar to potassium bisulfite and sodium bisulfite. Sulfuric acid has been used as a starch-modifier, a buffering and neutralizing agent, and a pH adjuster in the brewing industry.

The practice of exposing fruits to the poisonous fumes of sulfur was examined by Dr. Harvey W. Wiley, during the early days of the federal agency that later became the FDA. Dr. Wiley demonstrated that sulfur dioxide harmed the body, destroyed the fine flavor of dried fruit, and furthermore, was unnecessary. Dr. W. D. Bigelow, his assistant, developed a practical method of dipping the fruit into a harmless weak saline solution whereby the color and flavor were retained.

Dr. Wiley proposed a regulation to ban the use of

sulfur dioxide on dried fruits. A storm of protest followed from the fruit packers. Taking advantage of Wiley's temporary absence in Europe, they exerted pressure and had the regulation killed. As a result, sulfur dioxide has been permitted for use with many foods and beverages.

Bigelow's work, "Experiments Looking to Substitutes for Sulfur Dioxides in Drying Fruits," submitted in 1907, was denied USDA publication.

In 1911, the USDA instructed a group of "scientific experts" to investigate sulfur dioxide. The unpublicized report found that a portion of from three-tenths of a gram (the equivalent found in from six to eight ounces of dried fruit) up to one gram daily was sufficient to induce harm in some individuals if they ate such treated fruit over a period of months. Their symptoms included increased uric acid, albumin in the urine, destruction of blood corpuscles, anemia, belching of sulfur dioxide gas, teeth "on edge," inflammation of the mouth's mucous membrane, and malaise, headache, backache and nausea. Some individuals had the sensation of being cold, had dull eyes, and had a listless manner.

In 1929, a USDA official testified before Congress:

"Technology has developed, and it has gone, we believe, entirely too far. It has gone so far that the American dried fruit is now in very considerable disfavor abroad. In fact, England, France, Germany and Switzerland have placed partial embargoes on American dried fruits because of the excessive sulfur content. Japan has intimated that a complete embargo will be placed on American dried fruit. [We believe that] this enzyme [the cause of the darkening] can be controlled in other ways than by use of sulfur dioxide. If we can do that, it means the salvation of the whole dried fruit industry and particularly . . . where sulfur is so largely used."

Although little, if any, of the free sulfur dioxide may remain in the foodstuffs, their biological patterns have been altered, with a destruction of at least part of the valuable B-complex, thiamin. Frequently, "scientific research" has been cited to vindicate the use of sulfur dioxide on dried fruits. A study made in California (a state which leads in the sulfuring practice) demonstrated that sulfured fruit contained a higher vitamin C content than fruit dried without bleach or preservative. What the report failed to note was that vitamin C, present in the raw dried fruit, fed to laboratory animals, was lost in the cooked dried fruit consumed by man. Yet the cooking is necessary to eliminate a portion of the sulfur dioxide. One of the findings in the report is not usually quoted. A slight chalkiness and brittleness of the cutting teeth was observed as an abnormality in the animals fed the sulfured fruit.

Only certain portions of this report are generally quoted. For example, an article appeared in a leading medical journal assuring physicians that sulfured fruits are not harmful. Such a misstatement is important, since dried fruits are a favorite food for invalids and those on hospital diets. The limit of residue tolerance was incorrectly cited in the article as only a tenth of the actual amount generally used in commercial practice.

Although the use of sulfur dioxide as a preservative in dried fruits is under the jurisdiction of the Federal Food, Drug and Cosmetic Act, the FDA has never established an upper limit for its use. Despite the human toxicity of this compound, the agency grants it GRAS status, with this explanation: "In order to contain enough sulfur dioxide to be unsafe, a food would first become so distasteful as to be inedible, so its use is self-limiting." This position is arbitrary and unscientific. "Self-limiting" is a fatuous criterion for the use of sulfur dioxide or any other chemical in food. Additives should not be allowed in quantities up to the limit of consumer tolerance. *The point of harmfulness and toxicity may be reached at a far lower level than that which affronts the tongue or nose.*

Although FDA officials may not be worried about the toxic effects of sulfur dioxide and sulfites as food additives, officials from other quarters *are* concerned. A report issued by the Joint FAO/WHO Expert Committee on Food Additives in 1962, stated: "As the standard of diet improves, so the intake of sulfur dioxide rises. The affluent society consumes more wine, cider and beer, more soft drinks, fruit juices, fruit pulp, dried fruit and the many other good things that require the presence of this permitted preservative. Since wine alone may contain from 100 to 450 ppm of sulfur dioxide, it is clear that this one commodity can readily provide the full maximum acceptable intake of preservative." Even without wine in the diet, there is such a wide range of application for these sulfites that they may add up to a considerable daily consumption. In addition, sulfites in foods and beverages represent only one source of sulfites in the environment. The total body burden also includes sulfites as particulates in air pollution.

The question of human toxicity from the presence of sulfurous acid in wine, added as potassium pyrosulfite, was explored. Substantial quantities of the compound were detected in the wine. On an average, a total of two hundred milligrams per liter was added, of which up to fifty milligrams per liter was found to be present as free sulfurous acid.

Human tolerance to sulfurous acid varies considerably. When forty to fifty milligrams of free sulfurous acid was administered in wine to human volunteers, fifteen out of one hundred and fifty experienced gastric distress, increased salivation, headache, and diarrhea. Even as little as ten to twenty milligrams of sulfur dioxide in a 0.04 percent water solution led to nausea and headache in some human subjects.

Additional concern about the use of sulfites in food resulted from experiments with rats fed sodium bisulfite in the diet. The animals grew less rapidly, had stunted growth, suffered severe diarrhea and other adverse symptoms from thiamin deficiency. Thiamin-deficient diets were rendered more toxic by the addition of the sulfites. Questions have been raised regarding the ability of sulfites not only to lower the nutritional values of foods, but their possible interactions with calcium, vitamin A and other B vitamin fractions in addition to thiamin.

Xylitol: In 1963, the FDA approved xylitol for special dietary uses as a sugar substitute for diabetics in dietetic jams and jellies, as a sweetener in chewing gum, and in vitamin coatings. Xylitol is a sugar alcohol. (*See Sorbitol*). Like other sugar alcohols, xylitol, even in moderate doses, has a laxative effect in some persons. The sugar alcohols, as unabsorbed sugars, lead to osmotic diarrhea, are attacked by gut microflora, and cause flatulence and general intestinal disturbance.

By 1971, reports that injections of xylitol in humans had caused adverse reactions had prompted the agency to consider withdrawal of the limited use of xylitol. However the proposal to withdraw xylitol even from limited use was never finalized.

By 1976, xylitol was being hailed by Finnish researchers as a sweetener for chewing gum that did not promote dental decay, and even healed incipient tooth cavities. This feature prompted the National Institute for Dental Research to grant research funds, and American school children participated in an experiment to chew xylitol-containing chewing gum. The three-year experiment was abruptly halted, after the students had chewed only nine sticks each, over a three-day period, due to preliminary findings released from Great Britain. Results released from a mouse study showed that a substantial number of mice fed high doses of xylitol developed bladder stones late in their lives; some of these mice developed bladder tumors. In a rat study, although no bladder stones or tumors were found, some rats showed changes in their adrenal glands. In a few cases, these rats showed adrenal tumors. Since the studies were preliminary, the FDA awaited a final report. The FDA took no action to revoke approval for xylitol's use in special dietary foods. However, some manufacturers voluntarily withdrew xylitol from their products.

What Is To Be Done?

Vigorous enforcement of the Federal Food, Drug and Cosmetic Act is essential. As this book illustrates, by many examples, the FDA has a record of consistent and repeated failures to enforce regulations formulated to protect consumers.

The Delaney Clause, very clear in intent, is being flagrantly violated by the FDA's failure to act promptly and decisively when substances are found to be cancer-inciting. The Clause should not be weakened, but rather extended, to include mandatory tests for mutagenic, teratogenic, and other forms of subtle biological damage that chemical food additives are capable of inflicting.

The GRAS list must not be allowed to continue serving as a privileged status by exempting food additives from being adequately tested merely because they have been in long-time use. Up to the present, the expenses for safety evaluation tests have been the responsibility of the manufacturer. The FDA has brazenly proposed to transfer this financial burden to taxpayers. This must not be done. The cost of testing for profitable food chemicals must be borne by the food processors.

Excellent suggestions were presented by Dr. Epstein, during the congressional hearings on Chemicals and the Future of Man, regarding the need for legislation to ensure impartial and competent testing of all synthetic chemicals for which human exposure is anticipated. Since these remarks are applicable to food additives as well, they are worth considering. Dr. Epstein stated, in part: "The present system of direct, closed-contract negotiations between manufacturing industries and commercial and other testing laboratories is open to abuse, creates obvious mutual constraints, and is thus contrary to consumer, and long-term industrial interests. One possible remedy would be the introduction of a disinterested advisory group or agency to act as an intermediary between manufacturers and commercial and other testing laboratories . . .

"Additionally, the development of independent nonprofit research groups and centers concerned with problems of consumer, occupational, and environmental safety should be encouraged in an effort to develop constraint-free research, consultative and training centers. . . . Further consideration should be directed to the possible need for an independent consumer regulatory protection agency, and to the development of stronger scientific and legal representation of consumer, occupational and environmental public interest groups in all agencies concerned directly or indirectly in these areas.

"It is critical that industry, commercial testing, and other testing laboratories be provided with unambiguous guidelines as to exactly what tests are required and as to the necessary protocols for such tests. These protocols should relate to tests for both human and environmental safety and also delineate principles for interpretation. Concerned parties should be given ample opportunity to challenge or otherwise comment on such protocols before their promulgation is formally ratified. Such protocols should be subject to mandatory annual review.

"These procedures will provide a clearer definition as to the precise relative responsibilities of industry and regulatory agencies in safety evaluation, and thus minimize existing mutual constraints in these areas. Industry will thus know what tests are required, and hence be able to predetermine the approximate cost of such tests before deciding whether or not to proceed with further product testing and development.

"Further legislation concerning public access to data is critically needed. All formal discussions between agencies, industry, and expert governmental and nongovernmental committees on all issues relating to human safety and environmental quality, and all data relevant to such discussions properly belong to the public domain and should be a matter of open record. Such records, including clear statements, by all concerned, of possible conflicts of interest, should be immediately available to the scientific community, scientific and legal representatives of consumer, occupational, and environmental groups, and to other interested parties. . . .

"In addition to open access of data on all issues of public health and welfare, and environmental safety, it is essential that the interests of consumer, occupational, and environmental groups be adequately represented, legally and scientifically, at the earliest formal stages of such discussions. Decisions by agencies on technological innovations or on new synthetic chemicals after closed discussions on data which have been treated confidentially are unacceptable. Consumer, occupational and environmental safety apart, such decisions are contrary to the long-term interests of industry, which should be protected from perforce belated objections."

These recommendations of Dr. Epstein, if applied to food additives, would make it possible to have a far greater amount of consumer protection than presently exists.

At present, what can you, as a concerned consumer, do for yourself regarding food additives? You cannot completely avoid these substances, but you can minimize your intake. By now, it should be apparent to you that many of the food additives are used in highly processed convenience foods. Repeatedly, items such as baked goods, ice cream, sherbet, candies, confections, soft drinks, dessert powders, breakfast cereals, luncheon meats, frankfurters, sausages, snack foods, and others appear on lists in which many additives are used. *By eliminating convenience foods from your daily diet, you will succeed in reducing your total intake of chemical additives.* It is wise to avoid these foods for other reasons. They are depleted of vital nutrients, and are unreliable sources of minerals, vitamins, proteins, enzymes, and other essentials for maintaining good health.

Choose, as far as possible, foods close to their natural state: fresh vegetables and fruits, meats, fish, fowl, eggs, nuts, grains, natural cheeses, butter and crude vegetable oils. These are the foods that sustain human life.

Suggested Reading

Arena, Jay M., M.D., *Poisoning, Chemistry, Symptoms, Treatments*. Springfield, Ill.: Charles C. Thomas, 1963.

Bicknell, Franklin, M.D., *Chemicals in Food & in Farm Produce: Their Harmful Effects*. London: Faber & Faber, 1960; also New York: Emerson, 1960.

Furia, Thomas E., ed., *Handbook of Food Additives*. Cleveland, Ohio: The Chemical Rubber Co., 1968.

Hueper, Wilhelm C., M.D., Carcinogens in the Human Environment. *Archives of Pathology*, A.M.A. Vol. 71, March 1961, pp. 355-380; Vol. 71, April 1961, pp. 237-267.

Hueper, Wilhelm C., M.D., and Conway, W.D., *Chemical Carcinogenesis and Cancers*. Chicago: Charles Thomas, 1964.

Joint FAO/WHO Expert Committee on Food Additives:

—Specifications for identity and purity of food additives: Food Colors, Vol. 11, 1963.

—Specifications for the identity and purity of food additives and their toxicological evaluation: Food colors and some antimicrobials and antioxidants. FAO Nutr. Meet. Rep. Ser. No. 38, WHO Tech. Rep. No. 309, 1965.

—Specifications for identity and purity and toxicological evaluation of some food colors. FAO Nutr. Meet. Rep. Ser. No. 38B, WHO/Food Add/66.25, 1966.

—Specifications for the identity and purity of food additives and their toxicological evaluation: Some emulsifiers and stabilizers and certain other substances. FAO Nutr. Meet. Rep. Ser. No. 43; WHO Tech. Rep. Ser. No. 373, 1967.

—Specifications for the identity and purity of food additives and their toxicological evaluation; Some food colors, emulsifiers, stabilizers, anticaking agents, and certain other substances. FAO Nutr. Meet. Rep. Ser. No. 46, WHO Tech. Rep. Ser. No. 445, 1970a.

— Toxicological evaluation of some food colors, emulsifiers, stabilizers, anticaking agents and certain other substances. FAO Nutr. Meet. Rep. Ser. No. 46A; WHO Food Add/70.36, 1970b.

— Specifications for the identity and purity of some food colors, emulsifiers, stabilizers, anticaking agents and certain other substances. FAO Nutr. Meet. Rep. Ser. No. 46B; WHO/Food Add/70.37, 1970c.

Kraybill, H. F., Carcinogenesis associated with foods, food additives, food degradation products, and related dietary factors. *Clinical Pharmacology & Therapeutics*, Vol. 4, No. 1, January-February 1963, pp. 73-87.

Manufacturing Chemists' Association, *Everyday Facts About Food Additives*. Washington, D.C., no date.

Manufacturing Chemists' Association, *Food Additives: What they are/How they are used*. Washington, D.C., 1971.

National Academy of Sciences-National Research Council, *Chemicals Used in Food Processing*. Washington, D.C., Publication 1274, 1967.

National Academy of Sciences-National Research Council, *Food Colors*. Washington, D.C., 1971.

National Academy of Sciences-National Research Council, *Problems in the Evaluation of Carcinogenic Hazards from the Use of Food Additives*. Washington, D.C., Publication 749, 1959.

Randolph, Theron G., M.D., *Human Ecology and Susceptibility to the Chemical Environment*. Chicago: Charles Thomas, 1962.

Turner, James S., *The Chemical Feast*. New York: Grossman, 1970.

GOVERNMENT PUBLICATIONS

Dietary Nitrite in the Rat, Paul M. Newberne, Final Report on Contract FDA 74-2181, Department of Health, Education and Welfare, May 18, 1978.

Evaluation of the Health Aspects of Butylated Hydroxytoluene as a Food Ingredient, prepared for FDA by FASEB, U.S. Department of Commerce, National Technical Information Service, PB-259 917, July 1973.

Evaluation of the Health Aspects of Caffeine as a Food Ingredient, prepared for FDA by FASEB, U.S. Department of Commerce, National Technical Information Service, PB-283-441, 1978.

Evaluation of the Health Aspects of Sorbitol as a Food Ingredient, prepared for FDA by FASEB, U.S. Department of Commerce, National Technical Information Service, PB-221-951, December 1972.

Final Report on Nitrites and Nitrosamines, Report to the Secretary of Agriculture by the Expert Panel on Nitrites and Nitrosamines, Food Safety and Quality Service, U.S. Department of Agriculture, February 1978.

Hearings before the House Select Committee to Investigate the Use of Chemicals in Foods & Cosmetics. 81st Congress, 1st session, 1950; 82nd Congress, 1st session, Parts 1 & 2, 1951; and Hearings before a Subcommittee of the Committee on Interstate & Foreign Commerce, on Food Additives. 85th Congress, 1957-58, Parts 1, 2 and 3.

Hearings before the Committee on Interstate and Foreign Commerce, House or Representatives, on Color Additives. 86th Congress, 2nd session, 1960.

Hearings before the Subcommittee on Reorganization and International Organizations of the Committee on Government Operations, United States Senate, Interagency Co-ordination in Environmental Hazards (Pesticides). Parts 1-7, 1963, Part 8, 1964; and Appendix 1-6 to Part 1.

Hearings before the Subcommittee on Antitrust & Monopoly of the Committee on the Judiciary, United States Senate, on Packaging & Labeling Legislation. 88th Congress, 1st session, Parts 1-3, 1963-64.

Hearings before a Special Subcommittee of the Committee on Labor & Welfare, United States Senate, to Amend Section 402 (d) of the Federal Food, Drug & Cosmetic Act. 89th Congress, 1st session, 1965 (this hearing dealt with the subject of additives in candy).

Report of the Secretary's Commission on Pesticides and Their Relationship to Environmental Health, Parts 1 and 2, U.S. Department of Health, Education and Welfare, December 1969.

Hearings before a Subcommittee of the Committee on Government Operations, House of Representatives. 91st Congress, 2nd Session, June 10, 1970. Cyclamate Sweeteners; 36th Report by the Committee on Government Operations,

Oct. 8, 1970, Regulation of Cyclamate Sweeteners.

Hearings before a Subcommittee of the Committee on Government Operations, House of Representatives. 92nd Congress, 1st Session, March 29, 1971, Recall Procedures of the Food and Drug Administration; 1st Report by the Committee on Government Operations, October 21, 1971, Recall Procedures of the Food and Drug Administration.

Hearings before a Subcommittee of the Committee on Government Operations, House of Representatives. 92nd Congress, 1st Session, March 1971, Regulations of Food Additives and Medicated Animal Feeds.

Hearings before the Subcommittee on Executive Reorganization and Government Research of the Committee on Government Operations, United States Senate. 92nd Congress, 1st Session, April 1971, Chemicals and the Future of Man.

Hearings before the Subcommittee on Health and the Environment of the Committee on Interstate and Foreign Commerce, House of Representatives, 95th Congress, 1st Session, March 21 and 22, 1977, Proposed Saccharin Ban—Oversight.

Hearings before the Subcommittee on Health and Scientific Research of the Committee on Human Resources, United States Senate, 95th Congress, 1st Session, June 7, 1977, The Banning of Saccharin, 1977.

Report of the Comptroller General of the United States, Need to Resolve Safety Questions on Saccharin, HRD-76-156, August 16, 1976.

Saccharin: Technical Assessment of Risks and Benefits, Report No. 1, Committee for a Study on Saccharin and Food Safety Policy, Assembly of Life Sciences/Institute of Medicine, National Research Council/National Academy of Sciences, Washington, D.C., November 1978.

Safety and Suitability of Modified Starches for Use in Baby Foods, Subcommittee on Safety and Suitability of MSG and Other Substances in Baby Foods, Food Protection Committee, Food and Nutrition Board, National Research Council, Washington, D.C., September 1970.

Tentative Evaluation of the Health Aspects of Certain Glutamates As Food Ingredients, prepared for FDA by FASEB, FDA 223-75-2004, Life Sciences Research Office, Federation of American Societies For Expermental Biology, 1976.

PERIODICALS

Chemical & Engineering News, Washington, D.C.: American Chemical Society. "Food Additives" Part 1, October 10, 1966; Part 2, October 17, 1966; "Chemical Mutagens, the Road to Genetic Disaster?" Part 1, May 19, 1969; Part 2, June 2, 1969; "Food Processing: Growth in New Directions," August 23, 1971.

Consumer Bulletin. Washington, New Jersey: Consumers' Research, Inc. *Consumer Bulletin Annual*, Food and Nutrition Section. Washington, New Jersey; Consumers' Research, Inc.

Food Chemical News. Washington D.C.: Food Chemical News, Inc.

Food and Cosmetic Toxicology. Oxford, England: Pergamon Press.

Food Engineering. Philadelphia, Pennsylvania: Chilton Company.

ARTICLES

Tom Alexander, "The Hysteria About Food Additives," *Fortune*, March, 1972.

Arthur D. Colman, M.D., "Possible Psychiatric Reactions to Monosodium Glutamate," letter to the editor, *The New England Journal of Medicine*, Vol. 209, No. 16, October 19, 1978, page 902.

Mary G. Enig, Robert J. Munn and Mark Keeney, "Dietary Fat and Cancer Trends—a Critique," *Federation Proceedings*, Vol. 37, No. 9, July 1978, pages 2215-20.

G. O. Kermode, "Food Additives," *Scientific American*, March, 1972.

Jim Krueger, "MSG: One of Food Industry's Most Studied Ingredients," *Processed Prepared Food*, January 1979, pages 128-135; 140.

National Dairy Council, "Coronary Heart Disease: Risk Factors and the Diet Debate," submitted to the U.S. Select Committee on Nutrition and Human Needs, July 26, 1974.

Alan T. Spiher, Jr., J.D., "Food Ingredient Review: Where It Stands Now," *FDA Consumer*, June 1974, pages 23-26.
—"The GRAS List Review," *FDA Papers*, December 1970/January 1971.

Index

Acidifiers. *See* Acidulants; Buffers; Neutralizers
Acidifying agents. *See* Acidulants; Buffers; Neutralizers
Acidulants, 13; *see also* Buffers; Neutralizers
Acronym identification, 39-40
Acrylonitrile, 4
Additives, food
   animal tests on, 5-6, 8-9; *see also* Animal tests
   common, 2-3, 41-113
   consumption of, 2
   functions of, 13-34
   hazards of, 2
   interaction with other environmental factors, 4
   laboratory tests on, 6
   mutagenicity of, 10
   safe levels of, 9, 10
   safety evaluation of, 2-10
   synergism of, 6-7
   toxicity of, 6-7, 41
   types of intentional, 13-34
   see also specific additives
Adipose, definition of, 35
Adjuvants, flavoring, 14
Advisory Panel on Mutagenicity, 10
Aerating agents, 14
Albany Medical College, 56-57
Alcohol
   asbestos in, 99
   interaction of, with barbiturates, 7
   interaction of, with carbon tetrachloride, 7
Alkalies, 14; *see also* Buffers; Neutralizers
Allergic reactions to
   aspirin, 66
   benzoates, 66
   BHT, 45
   MSG, 83-84, 88
   propionates, 52-53
   tartrazine, 65-66
Allura. *See* FD&C Red No. *40*
AMA, 30
AMA Council on Foods and Nutrition, 85
Amaranth, 63
American Academy of Pediatrics, 30
American Medical Association. *See* AMA
Aminotriazole, 9
Anderson, Thomas A., 80
Animal tests on
   additives, 5-6, 8-9
   azo dyes, 59
   benzoic acid, 42-43
   BHA, 43
   BHT, 44-46
   Blue No. *1*, 66-67
   Blue No. *2*, 67
   caffeine, 49
   calcium disodium EDTA, 51
   carrageenan, 54-57
   food flavorings, 70-72
   Green No. *3*, 66
   modified food starches, 80-81
   MSG, 85, 87, 89
   Orange B, 65
   Red No. *40*, 62-63
   saccharin, 92
   sodium carboxymethylcellulose, 101
   sulfur dioxide, 110, 112
   xylitol, 113
Antibiotics, 14-15
Antibrowning agents, 15
Anticaking agents 15. *See* Silicates
Antifirming agents. *See* Antistaling agents
Antifoaming agents. *See* Defoaming agents

Antimold agents, 15; see also Preservatives
Antimycotic agents. See Antimold agents; Antirope agents
Antioxidants, 15-16
  MSG as, 83
  synthetic, 16
  see also BHA; BHT; Preservatives
Antirope agents, 16
Antistaling agents, 16-17; see also Dough conditioners
Antisticking agents. See Release agents
Anxiety neurosis, and caffeine, 50
Arthritic symptoms, and nitrates, 102
Artificial colors. See Colors, food
Artificial sweeteners. See Nonnutritive sweeteners
Asbestos
  interaction of, with cigarette smoke, 7
  in silicates, 97-99
Ascorbates, 104, 106
Ascorbic acid. See Ascorbates
Aspirin, allergy to, 66
Ataxia, 35
Azo dyes, 59; see also FD&C Citrus Red 2; Orange B

Baby food
  carrageenan in, 54
  MSG in, 85-87
  starch in, 79-81
Barbiturates, interaction with alcohol, 7
Base components, chewing gum, 17
Benzine, 8
Benzoate of soda, 41-43
Benzoates, allergy to, 66
Benzoic acid, 41-43
  animal tests on, 42
  in berries, 42
  synergism with sodium dioxide, 43
  toxicity of, 42

Beta-naphthylamine
  animal tests on, 5
  as carcinogen, 8
  in Orange B, 65
BHA, (Butylated hydroxyanisole), 43-45
  animal tests on, 43
  safety tests on, 44
  use of, 43
BHT (Butylated Hydroxytoluene), 44-46
  animal tests on, 44-46
  petition for ban of, 46
  toxicity of, 45
  use of, 44
Bicknell, M.D., Franklin (*Vitamins in Medicine*), 74-75
Bigelow, Dr. W. D., 108, 109
Binders, 17
Biochemical equilibrium, and caffeine, 50
Birth defects and
  carrageenan, 56
  chemical additives, 2
  see also Chromosomal damage; Genetic damage; Teratogen
Bleaching agents, 17-18; see also Bread improvers; Dough conditioners; Maturing agents
Bodying agents, 18
Brain and
  cola beverages, 49
  erythrosine, 62
  MSG, 85, 87, 89-90
Bread emulsifiers. See Antistaling agents
Bread improvers, 18; see also Bleaching agents; Dough Conditioners; Maturing agents
Bread, preservatives in, 52-53; see also BHA; BHT
Brilliant Blue FCF. See FD&C Blue No. *1*
Britain
  BHT ban in, 44-45
  and Blue No. 2, 67
  and Citrus Red 2, 64
  food color report of, 60

Bryan, M.D., George, 92
Buffering agents. *See* Buffers
Buffers, 19; *see also* Acidulants; Alkalies
Butylated hydroxyanisole. *See* BHA
Butylated hydroxytoluene. *See* BHT
B vitamins, and sulfur dioxide, 110, 112

Caffeine, 46-50
   animal tests on, 47
   effects of, 47-50
   children's consumption of, 48-49
   in cola drinks, 47-50
   teratogenicity of, 47, 49
   toxicity of, 47-48
Calcium, assimilation of, 53
Calcium disodium EDTA, 50-52
   animal tests on, 51
   effects of, 51-52
   use of, 50-51
Calcium propionate, 52-53
   and calcium assimilation, 53
   effect of, 52-53
   use of, 52-53
Calorie Control Council, 94
Canada and
   asbestos, 99
   nitrates, 103
   fats and oils, 77
   Red No. *4*, 62-63
   saccharin, 93
Canadian Dental Association, 93
Canadian Diabetic Association, 93
Canadian Food and Drug Directorate, 62, 77, 93, 99, 103
Canadian Health and Welfare, 77
Canadian Medical Association, 93
Canadian National Health Protection Branch, 93
Cancer
   and asbestos, 99
   and Blue No. *1*, 66-67
   and Blue No. 2, 67
   and chemical additives, 2
   and dietary fat, 76
   fear of, 9-10
   and nitrites/nitrates, 103, 104, 105, 106
   and saccharin, 92, 93, 94, 95-96
   and sodium carboxymethylcellulose, 101
   see also Carcinogen; Cocarcinogen
Canned food, MSG in, 82-83
Canine hysteria, 18
Carbon tetrachloride, interaction with alcohol, 7
Carboxymethylcellulose. *See* Sodium carboxymethylcellulose
Carcinogen
   caffeine as, 50
   carrageenan as, 57
   chemicals as, 7
   definition of, 35
   epichlorohydrin as, 81
   Orange B as, 65
   Red No. *40* as, 63-64
   relation of, to toxic properties, 7-8
   triphenylmethane coal tar colors as, 59
   wood smoke as, 31
   see also Cancer; Cocarcinogen
Carcinoma, 35
Carrageenan, 53-58
   animal tests on, 54-57
   and cancer, 57
   hazards of, 54-57
   properties of, 53-54
   toxicity of, 54
   use of, 53, 57-58
Carriers, 19
Casein, 33
Cecum, 35
Cellulose gum. *See* Sodium carboxymethylcellulose
Certified Color Industry Committee, 60-61
Certified colors, 58-69
   coal tar dyes, 59-61
   description of, 20-21
   FD&C Blue No. *1*, 66-67
   FD&C Blue No. 2, 67-68

FD&C Green No. 3, 66, 68
FD&C Red No. 2, 63
FD&C Red No. 3, 61-62
FD&C Red No. 4, 62, 68
FD&C Red No. 40, 62-64
FD&C Yellow No. 5, 65-66, 69
FD&C Yellow No. 6, 65, 68, 69
natural sources for, 67
Orange B, 65
Cheese, bleaching agents in, 18
Chelating agents. See Sequestrants
Chelators. See Sequestrants
Chemicals and the Future of Man hearing, 115
Chewing gum
base components in, 17
xylitol in, 112-113
Chickens, 9
Children
behavioral effects of caffeine on, 48-50
hyperactivity of, 62
and MSG, 88-89
and sorbitol, 107
see also Baby food; Birth defects, Infants
Chinese Restaurant Syndrome, 84-85, 88-89, 90
Chocolate
carrageenan use in, 57
emulsifiers in, 23
see also Caffeine
Chromosomal damage, and caffeine, 50; see also Birth defects; Genetic damage; Teratogen
Cholesterol and
fatty acids, 76
modified starches, 80
Cigarette smoke
interaction of, with asbestos, 7
interaction with radiation, 7
Clarifiers. See Clarifying agents
Clarifying agents, 20
Clouding agents, 20
Coal tar dyes. See Certified colors
Coating agents, 20

Cocarcinogen
chemicals as, 7
definition of, 35
see also Cancer; Carcinogen
Coffee. See Caffeine
Cola beverages. See Caffeine
Collagen, 35
Collagen granuloma, 56
Colitis, ulcerative, 55, 56
Colman, Arthur, M.D., 90
Color Additives Amendment of 1960, 9, 61
Colors, food. See Certified colors
Commission of the Health Association (France), 91
Congressional Hearings on Regulations of Food Additives and Medicated Animal Feeds, 104
Consumers' Research Bulletin, 84
Convenience foods
flavoring additives in, 72-73
hazards of, 118
Convulsions, 37
Corn oil, 77
Coumarin, 70
Cranberries, 9
Crisping agents. See Firming agents, 21
Crumb-firming agents. See Antistaling agents
Crystalline inhibitors, 21
Crystallization modifiers, 21
Curing agents, 21; see also Fixatives; Sodium Nitrite/nitrate
Cyclamates
ban on, 92
human tests of, 9
inadequate testing of, 11
toxicity of, 6

Defoaming agents, 21; see also Surfactants
Delaney Clause
and carcinogens, 9-10, 103
and nitrites/nitrates, 105

and saccharin, 93-96
violation of, 115
Delaney Hearings, 9
Deoxyribonucleic acid. *See* DNA
Diabetics, and sorbitol, 106
Diarrhea
carrageenan as cure for, 53
and sorbitol, 107
and sulfur dioxide, 112
and xylitol, 112
Disease, and lack of essential fatty acids, 75, 76
DNA
definition of, 35
effect of caffeine on, 47
Dopamine, 62
Dough conditioners, 22; *see also* Antistaling agents; Bleaching agents; Bread improvers; Maturing agents
Dried fruits. *See* Fruits
Dusting agents. *See* Release agents
Dyes, food. *See* Certified colors
Dynapol, 6

Edema, 35
Emulsifiers, 22-23; *see also* Stabilizers
Emulsifying agents. *See* Emulsifiers
Enzymes, 23
Epichlorohydrin, 81
Epstein, Samuel S., M.D., 8-9, 10, 103, 115-117
Essential fatty acids (EFA)
definition of, 35
destruction of, in hydrogenation, 74-75
See also Fatty acids
Essential oils, toxicity of, 72
Erythrosine. *See* FD&C Red No. 3
Escherichia coli and
caffeine, 47
erythrosine, 61
Extenders, 23; *see also* Fillers

FAO/WHO (Food and Agricultural Organization/World Health Organization)
Blue No. *1*, 67
Blue No. *2*, 67, 68
Citrus Red *2*, 64
Red dye No. *40*, 62
sulfur dioxide, 111
unreliability of animal tests, 5-6
xanthene dye, 61
FASEB (Federation of American Societies for Experimental Biology), 5
and BHA, 43-44
and BHT, 45-46
and MSG, 89
Fast Green FCF. *See* FD&C Green No. *3*
Fat, hydrogenated. *See* Hydrogenated fat
Fatigue, 50
Fatty acids
and caffeine, 47
see also Essential fatty acids
FDA
antibiotics, ban of, 15
and BHT, 46
and Blue No. *1*, 67
and Blue No. *2*, 67
and carrageenan, 56-57
and Citrus Red *2*, 64
and cola drinks, 47, 48
and consumer protection, 115
and fats and oils, 77
and food and drug dyes, 60-61, 67, 68
and food flavorings, 71-73
and modified food starch, 80
and MSG, 88, 89, 90, 91
and nitrites/nitrates, 104, 105, 106
and Orange B, 65
public access to files of, 41
and Red No. *2*, 63
and Red No. *40*, 62-63
and saccharin, 92-95, 97

safety testing of, 2-4, 6, 11
and sodium carboxymethylcellulose, 101
and sorbitol, 106
and sulfur dioxide, 108, 110-111
and vitamin D in milk, 30
and xylitol, 112-113
and Yellow No. 5, 65-66
FD&C Blue No. 1, 66-67
FD&C Blue No. 2, 67-68
FD&C Citrus Red No. 2, 7, 64
FD&C Green No. 3, 59, 66, 68
FD&C lakes, 27
FD&C Red No. 2, 7, 64
FD&C Red No. 3, 61-62
FD&C Red No. 4, 62, 68
FD&C Red No. 40, 62-64
FD&C Yellow No. 5, 65-66, 69
FD&C Yellow No. 6, 65, 68, 69
Federal Food, Drug and Cosmetic Act, 115
Federal Food Additive Amendment of 1958, 2, 9, 70
Federal Register, 65
Federation of American Societies for Experimental Biology. *See* FASEB
Fibrosis, 35
Fillers, 24
Fining agents. *See* Clarifying agents
Firming agents, 24
Fixatives, 24
Flavor enhancers, 24-25; *see also* MSG
Flavor enzymes, 25
Flavoring Extract Manufacturers Association (FEMA), 70, 71
Flavorings, artificial, 25; *see also* Flavorings, food
Flavorings, food, 25-26, 69-73
description of, 25-26
essential oils, 72
hazards of, 72
toxicity tests on, 71-72
U.S. consumption of, 72-73
see also Adjuvants, flavoring; Flavor enhancers; Flavor enzymes; Flavoring, artificial; Flavoring, imitation; Oleoresins
Flavorings, imitation, 26; *see also* Flavorings, food
Flavorings, natural. *See* Flavorings, food
Flavorings, synthetic. *See* Flavorings, food
Flavor intensifiers. *See* Flavor enhancers
Flavor modifiers. *See* Flavor enhancers
Flavor potentiators. *See* Flavor enhancers
Flour
bleaching of, 17-18
enrichment of, 30
see also Bread
Foaming agents. *See* Aerating agents
Food, canned, 82-83
Food, convenience, 72-73
Food, Drugs, & Cosmetics Colors. *See* Certified colors
Food, low-calories. *See* low-calorie foods
Food additives. *See* Additives, food; specific additives
Food and Agricultural Organization/World Health Organization. *See* FAO/WHO
Food and Drug Administration. *See* FDA
Food and Drug Research Labs, Inc., 56
Food colors. *See* Certified colors
Food dyes. *See* Certified colors
Food flavorings. *See* Flavorings, food
Food processors
and FDA safety rules, 2-3
and safety testing, 115-117
Food Protection Committee, Food & Nutrition Board. *See* NAS-NRC
France, saccharin ban in, 91
French Academy of Medicine, 69

Freshness preservers. *See* Antioxidants
Fruits and sulfur dioxide, 108-109
Fumigants, 27
Fungi, 52
Fungicides, 27

Gastric ulcers, and carrageenan, 55
Gastrointenstinal cancer, and asbestos, 99
Gastrointestinal distress, and propionates, 52-53
General Accounting Office, and carrageenan, 57
Generally Recognized As Safe list. *See* GRAS list
Genetic damage
 and additives, 2
 see also Birth defects; Chromosomal damage;. Teratogen
German Research Institute for Food Chemistry, 61
Germany, saccharin ban in, 91
Glazing agents, 27; *see also* Coating agents; Release agents
Glossary, medical term, 35-36
Grain
 bleaching of, 17-18
 enrichment of, 30
 see also Bread
Granuloma, 36-56
GRAS list
 benzoic acid on, 41-43
 BHA on, 43
 BHT on, 44, 45
 caffeine on, 46, 48
 calcium propionate on, 52-53
 carrageenan on, 56
 food flavorings on, 70, 71, 72
 misuse of, 115
 MSG on, 82, 84, 86, 88-89, 91
 and safety testing, 2-3
 sodium carboxymethylcellulose on, 101
 sulfur dioxide on, 110
 vitamine D fortification on, 30
Great Britain. *See* Britain

Gum, chewing. *See* Chewing gum

Hazeltine Labs, 63
Heinz bodies, 36, 59
Heinz (H.J.) Company, 80
Hippuric acid, 42
Heuper, Wilhelm C., M.D., 101-102
Humectants, 27
Hungary, saccharin ban in, 91
Hydrogenated fat, 73-78
 and disease, 76
 effect of, on EFA, 73-76
 method of producing, 74, 77
 and saturation, 77
Hydrolyzed vegetable protein, 86
Hydroscopic agents. *See* Humectants
Hydrous magnesium silicate. *See* Silicates
Hyperactivity and
 erythrosine, 62
 MSG, 90
Hyperplasia, 36
Hypothalmic brain lesions, 89
Hysteria, canine, 18

IARC/WHO (International Agency for Research on Cancer/World Health Organization), and carrageenan, 57
Indigo Carmine. *See* FD&C Blue No. 2
Infants
 and vitamin D, 30
 see also Baby food; Birth defects; Children
Inflammation, and carrageenan, 55
Ingotine. *See* FD&C Blue No. 2
International Agency for Research on Cancer/World Health Organization. *See* IARC/WHO
International Glutamate Technical Committee, 90
Institute of Food Technologists, 68

Iodine levels, and xanthene dye, 61
Irish moss. *See* Carrageenan
ITT Continental Baking Company, 66

Jasmin, G., M.D., 101
Jellying. *See* Carrageenan
Journal of the American Medical Association, and nitrosamines, 104

Keeney, Dr. Mark, 76
Kennedy, Donald, 66
Keys, Ancel, 75-76
Khan, Paul, 66
Kidney damage and
  Blue No. 2, 67-68
  calcium disodium EDTA, 51
Kwok, Robert Ho Man, 84

Labeling of
  fats and oils, 77
  food colorings, 66
  food flavorings, 73
  MSG, 41, 84
  nonnutritive additives, 101
  saccharin, 94-95
Lancet, 76
Leavening agents, 28; *see also* Dough conditioners
Lecithin, 22
Lehman, Arnold J., M.D., 6
Lijinsky, Dr. William, 103
Low-calorie foods
  carrageenan in, 57-58
  sorbitol in, 106
  see also Nonnutritive sweeteners
Lubricating agents. *See* Release agents
Lung cancer, causes of, 7

McGovern Committee Hearings on Nutrition and Human Needs, 80
McGovern, Senator George, 3
Mannitol, 78

Margarine, 77-78
Massachusetts Institute of Technology, 104, 105
Mattson, Dr. Howard, 68
Maturing agents, 28; *see also* Bleaching agents; Bread improvers; Dough conditioners
Mayer, Dr. Jean, 86
Meclizine, 5
Medical terms, 35-36
Medical Tribune, 105-106
Mellanby, Sir Edward, 18
Merliss, Reuben R., M.D., 98
Metabolism, 36
Metal scavengers. *See* Sequestrants
Methionine sulfoxamine, 18
Miller, Dr. Sanford, 73
Mineral(s)
  effect of calcium disodium EDTA on, 51-52
  supplements. *See* Nutrient supplements
Modified food starches, 78-81
  in baby food, 79-80, 81
  chemical modification of, 80
  economy of, 79-80
  use of, 78-79
Moisture-retaining agents. *See* Humectants
Mold inhibitors. *See* Antimold agents
Mold retarders. *See* antimold agents
Monosodium glutamate. *See* MSG
MSG (Monosodium glutamate), 81-91
  and allergies, 83-84, 88
  animal tests on, 85, 87, 89
  antioxidative properties of, 83
  in baby food, 85-86, 87
  and Chinese Restaurant Syndrome, 83-84, 88-89, 90
  economy of, 83
  effect of, on humans, 86-87
  labeling of, 41, 84
  and psychological disturbances, 90

toxicity of, 88-89
use in U.S., 82
Mucilagen. *See* Carrageenan
Mutagen
  caffeine as, 47, 49, 50
  chemical additives as, 10
  definition of, 36
  xanthene as, 59, 61
Mutation, 36
Mycoses, 52

NAS-NRC
  and carrageenan, 56
  Food Protection Committee of, 3-4
  and saccharine, 93, 95-96
National Academy of Sciences, and modified food starches, 81
National Academy of Sciences-National Research Council. *See* NAS-NRC
National Cancer Institute. *See* NCI
National Institute for Dental Research, and xylitol, 112-113
National Institute of Environmental Health Sciences, 10
National Institute of Neurological and Communicative Disorders and Stroke, 61-62
Natural colors. *See* Certified colors, natural sources for
NCI (National Cancer Institute) and
  Red No. *40*, 63
  saccharin, 95
  sodium carboxymethylcellulose, 101
Necrosis, 36
Nervous system, caffeine and, 47-48
Neutralizers, 29; *see also* Acidulants; Alkalies
Neutralizing agents. *See* Neutralizers
Newberne, Dr. Paul, 105
New England Journal of Medicine, 84

Nitrilotriacetic acid, 11
Nitrites/nitrates. *See* Sodium nitrites/nitrates
Nitrogen trichloride, 18
Nitrosamines, 103-105
Nonnutritive additives. *See* Sodium carboxymethylcellulose
Nonnutritive sweeteners, 29-30; *see also* Cyclamates; Saccharin
Nutrient supplements, 30

Office of Technology Assessment (OTA), and saccharin, 94
Oils
  clouding agents in, 20
  corn, 77
  essential, 72
  see also Hydrogenated fats
Oleoresins, 30; *see also* Flavorings, food
Olney, John, 88
Orange B, 65
Oriental food, MSG in, 81-84, 88-90
Oxygen interceptors. *See* Antioxidants

Papilloma, 36
Paprika, 30
Para-cresidine, 63
Peritoneum, 36
pH, food, 13, 14
pH adjusting agents. *See* Alkalies
Plasticizers, 31
Polishing agents, 31; *see also* Coating agents
Polymer dyes, 69
Polyp, 36
Portugal, saccharin ban in, 91
Preservatives
  description of, 31
  see also Acidulants; Antibiotics; Antibrowning agents; Antimold agents; Antioxidants; Benzoic acid; BHA; BHT; Calcium disodium EDTA; Calcium propionate; Fumigants; Fungicides; Sequestrants

Propellants, 31-32; *see also* Aerating agents
Propionic acids, 52-53
Protein hydrolysate, 86
Psychological disturbances, and MSG, 90
Pyrazolone dye, 65-66

Radiation, interaction with cigarette smoke, 7
Rare sugars, 78, 106-107, 112-113
Registrars of Pharmacy, 93
Release agents, 22; *see also* Coating agents
Rest, need for, 50
Restaurants, MSG use in, 83, 84
Ribicoff, Senator Abraham, 1, 8, 10
Rice, asbestos in, 98
Rief-Lehrer, Liane, M.D., 88-89
Robbins, Frederick, M.D., 96
Romania, tests on BHT in, 44
Roosevelt, President Theodore, 91
Root beer, 70
Rope inhibitors. *See* Antirope agents
Rubicoff, Abraham, 1, 4, 8, 10

Saccharin, 91-97
  and cancer, 92, 93, 94, 95-96
  harmful effects of, 4, 91-92
  FDA ban on, 93-95, 97
  labeling of, 94-95
  safety tests on, 29, 92-93
  and tumors, 92-93
  use of, 91
Saccharin Study and Labeling Act, 94-95
Sackler, Arthur M., M.D., 105-106
Safrole, 70
Sarcomas
  and azo dyes, 59
  and carrageenan, 56
  definition of, 36
  and Green No. 3, 66

Saturated fat, 77
SCOGS and
  caffeine, 48-50
  modified food starch, 81
  MSG, 86-88, 89, 90
  sorbitol, 107
Seaweed. *See* Carrageenan
Select Committee on GRAS Substances. *See* SCOGS
Senate Subcommittee on Health and Scientific Research, 94, 95
Sequestering agents. *See* Sequestrants
Sequestrants
  description of, 32
  see also Calcium disodium EDTA
Sexual development, and MSG, 88
Skin ailments, and essential oils, 72
Silicates, 97
  asbestos, 97
  and cancer, 99
  forms of, 97
  talc, 97-98
  use of, 97
Siu, Dr. G.H., 49-50
Smoke. *See* Cigarette smoke; Wood smoke
Sodium bisulfite, combined with benzoic acid, 43
Sodium carboxymethylcellulose, 100-102
  animal tests on, 101
  and cancer, 101
  as stabilizer, 100
  use of, 100
Sodium nitrite/nitrate, 102-106
  animal tests on, 102-103
  and cancer, 103, 104, 105, 106
  and nitrosamines, 103-104
  safety of, questioned, 4
  toxicity of, 102, 103
  use of, 102
  and vitamin A, 102-103
Sodium propionate, 52-53

Softeners. *See* Plasticizers
Solvents, 32
Sorbitol, 106-107
   calories in, 106
   effect of, 107
   use of, 106-107
Soviet Union, test of benzoic acid, 42-43
Spain, saccharin ban in, 91
Sperm output, and caffeine, 49
Spices, and oleoresins, 30
Spoilage retarders, 13, 14-16
Stabilizers, 32-33; *see also* Sequestrants; Sodium carboxymethylcellulose
Stabilizing agents. *See* Stabilizers
Starches, modified food, 78-81
Stilbestrol, 9
Sugar alcohol. *See* Mannitol; Sorbitol; Xylitol
Sugar substitutes. *See* Mannitol; Nonnutritive sweeteners; Sorbitol; Xylitol
Sulfites. *See* Sulfur dioxide
Sulfur dioxide, 108-112
   animal tests on, 110, 112
   effects of, 109, 111-112
   toxicity of, 111-112
   use of, on fruits, 108-109
Sulfurous acid, 111-112
Sunset Yellow FCF. *See* FD&C Yellow No. 6, 65
Supplements, vitamin and mineral. *See* Nutrient supplements
Surface active agents. *See* Emulsifiers
Suspending agents. *See* Stabilizers
Surfactants, 33; *see also* Defoaming agents; Emulsifiers; Foaming agents
Synergism and
   caffeine, 50
   chemical additives, 6-7
Synthetic colors. *See* Certified colors
Synthetic sweeteners. *See* Nonnutritive sweeteners

Tachycardia
   and caffeine, 47
   definition of, 36
Talc, 97-98
Tartness, 13
Tartrazine. *See* FD&C Yellow No. 5, 65-66
Tea. *See* Caffeine
Teratogen
   caffeine as, 47, 50
   definition of, 36
   meclizine as, 5
   see also Birth defects Chromosomal damage; Genetic damage
Texturizers, 33
Thalidomide, 5
Thiamin, and sulfur dioxide, 110 112
Thickeners, 33. *See* Emulsifiers; Stabilizers; Vegetable gum
Thickening agents. *See* Thickeners
Tonic-clonic convulsions, 37
Toxic gases, 27
Toxicity
   additives, 6-7
   benzoic acid, 42-43
   BHT, 45
   caffeine, 47-48
   carrageenan, 54
   food colorings, 59, 60
   food flavorings, 71-72
   MSG, 88-89
   nitrites/nitrates, 102
   relation to carcinogenicity, 7-8
   of synthetic antioxidants, 16
Trace minerals, 51-52
Triphenylmethane dye. *See* FD&C Green No. 3
Tuberculosis, carrageenan and, 53
Tumors and
   Blue No. 1, 66-67
   Blue No. 2, 67
   carrageenan, 56

saccharin, 92-93
Turmeric, oleoresins and, 30

Ulceration, and carrageenan, 54-55, 56
Ulcerative colitis. *See* Colitis, ulcerative
Ulcers, gastric
  and caffeine, 47
  and carrageenan, 55
United Kingdom. *See* Britain
United States
  food flavorings use in, 69, 72-73
  overprocessing of food in, 78-79
  packaged food, consumption of, 1
  use of MSG in, 82
United States Department of Agriculture (USDA)
  inadequate safety testing by, 10-11
  and nitrites/nitrates, 104, 106
  and sulfur dioxide, 109
United States Public Health Service (USPHS), and nonnutritive additives, 100
Unsaturated fats. *See* Essential fatty acids; Oils
Urticaria, 36

Vegetable gum, 33-34; *see also* Stabilizers
Vinyl chloride, 4
Viscosity agents. *See* Emulsifiers
Vitamin A and
  nitrites/nitrates, 102-103
  sulfur dioxide, 112

Vitamin B complex, and sulfur dioxide, 110, 112
Vitamin C, and sulfur dioxide, 110
Vitamin D, in milk, 30
Vitamins in Medicine (Bicknell), 74-75
Vitamin supplements. *See* Nutrient supplements

Washing agents, 34
Water, asbestos in, 99
Water-retaining agents. *See* Humectants
Weak carcinogen. *See* Cocarcinogen
Wenzel, Duane G., 76
Wetting agents. *See* Emulsifiers
Whipping agents. *See* Aerating agents
Wiley, Harvey, M.D., and
  nitrites/nitrates, 102
  saccharin, 91
  sulfur dioxide, 108-109
Wine, and sulfurous acid, 111
Wood smoke, 31
World Health Organization. *See* FAO/WHO; IARC/WHO

Xanthene coal tar colors, 59, 61
Xylitol, 112-113
  in gum, 112
  as sugar alcohol, 106
  tests of, 112-113

Yeast foods, 34